Routledge Revivals

Population Geography: Progress and Prospect

First published in 1986, this book presents a comprehensive overview of the contemporary state of knowledge in the field of population geography. It discusses the contemporary state of the art and surveys new research developments and new thinking in the major branches of the subject. It thereby provides an introductory guide to contemporary trends and forms a reference point for future development in the subject.

Population Geography: Progress and Prospect

Edited by
Michael Pacione

Routledge
Taylor & Francis Group

First published in 1986
by Croom Helm

This edition first published in 2011 by Routledge
2 Park Square, Milton Park, Abingdon, Oxon, OX14 4RN

Simultaneously published in the USA and Canada
by Routledge
270 Madison Avenue, New York, NY 10016

Routledge is an imprint of the Taylor & Francis Group, an informa business

Publisher's Note

The publisher has gone to great lengths to ensure the quality of this reprint but points out that some imperfections in the original copies may be apparent.

Disclaimer

The publisher has made every effort to trace copyright holders and welcomes correspondence from those they have been unable to contact.

A Library of Congress record exists under LC Control Number: 86006210

ISBN 13: 978-0-415-61611-9 (hbk)
ISBN 13: 978-0-415-61612-6 (pbk)

Population Geography: Progress & Prospect

Edited by
MICHAEL PACIONE

CROOM HELM
London • Sydney • Dover, New Hampshire

Croom Helm Ltd, Provident House, Burrell Row,
Beckenham, Kent BR3 1AT

Croom Helm Australia Pty Ltd, Suite 4, 6th Floor,
64-76 Kippax Street, Surry Hills, NSW 2010, Australia

British Library Cataloguing in Publication Data

Population geography: progress and prospect.
1. Population geography
I. Pacione, Michael
304.6 HB1951

ISBN 0-7099-4045-9

Croom Helm, 51 Washington Street, Dover,
New Hampshire 03820, USA

Library of Congress Cataloging in Publication Data

Population geography.
(Croom Helm Progress in geography series)
Includes index.
1. Population geography. 2. Census. 3. Fertility, Human. 4. Mortality. 5. Population policy. I. Pacione, Michael. II. Series.
HB1951.P653 1986 304.6 86-6210
ISBN 0-7099-4045-9

Printed and bound in Great Britain by
Biddles Ltd, Guildford and King's Lynn

CONTENTS

TABLES AND FIGURES

Tables

Figures

To Christine, Michael John and Emma Victoria

PREFACE

The study of population represents an integral component of human geography, yet only since the early 1950s has population geography emerged from the academic shadows to become an acknowledged sub-branch of the discipline. While population is also a focus of investigation for other subjects, such as demography, economics and sociology, population geography is characterised by its particular perspective on spatial aspects of population and on the links between population and the environment. Contemporary investigations in population geography focus on the three major demographic factors of fertility, mortality and migration. Recent developments in population geography have mirrored trends in the discipline as a whole. These have included a shift of emphasis from description of patterns of population distribution and comparison per se, towards an emphasis on the processes underlying observed patterns. There has also been a re-ordering of research emphasis among the three main themes with greater attention devoted to population mobility and migration studies. This reflects both the important social and economic implications of population movements, and the availability of spatially disaggregated data. These changes have been accompanied by development of an applied or problem-oriented perspective, as demonstrated by policy-relevant investigations in the fields of settlement and economic planning and in the evaluation of the effects of population policies. Modern population geography is a multi-faceted subject area which seeks to blend traditional descriptions of fertility, mortality and migration patterns with analyses of the dynamic processes underlying these patterns.

This collection of original essays is designed to encapsulate the major themes and recent developments in a number of areas of central importance in population geography. The volume is a direct response to the need for a text which reviews the progress and current state of the subject and which provides a reference point for future developments in population geography.

Michael Pacione
University of Strathclyde
Glasgow

INTRODUCTION

The pluralistic nature of population geography, with its strong links to demography and population studies, is acknowledged in Chapter 1 in which Bob Woods reviews recent developments in the theory and methodology of the subject. The fundamental question of the definition of population geography is considered and a distinction is drawn between broad definitions (such as that employed by Trewartha) and narrower statements which seek to concentrate attention on demographic geography (ie. the study of variations in population dynamics via analysis of mortality, fertility and migration components at various scales.) Contemporary population geography accommodates researchers espousing both demographic as well as traditional geographic perspectives. While such plurality may promote progress in the subject it can also be argued that the failure to define a precise area of interest and expertise has contributed to the limited influence of population geography in the general field of population research. Attention then turns to theoretical considerations. The purpose of theory and the related question of how a theory can be validated are subject to detailed examination. The discussion then examines the different levels at which theories may be constructed. Three distinct scales are recognised. Micro theory, which seeks to understand individual behaviour, is illustrated with reference to the study of marriage practices within an environment of constrained choice. Middle-range theory which is concerned with the characteristics of groups of people is the most commonly employed level of conceptualisation. Work at this scale is exemplified by Easterlin's attempts to combine economic and sociological theories of fertility. Grand theory is directed at long term social, economic and demographic change and is best illustrated by demographic transition theory. The basis of a flexible theory of demographic behaviour based on the interaction of a variety of causal factors is suggested. Five principal methodologies for population geography are then considered. Empiricism has been the dominant approach to pattern identification and modelling with significant contributions being made in fields such as multiregional demographic accounting, and automated cartography. When attempting to establish causal relationships some form of positivism has normally been invoked. Though of continued relevance in population geography, the limitations of positivism have led to the consideration of alternative approaches, with for example, a behavioural perspective being of particular utility in migration studies. Despite several problems (such as the non-replicability of survey results) the behavioural and humanistic approaches can improve understanding of the forces operating to produce observed population patterns. The structural perspective views human behaviour as conditioned by the dominant mode of production, with each mode having its own laws of population. The potential and problems of each of these methodologies are assessed. Two fundamental issues for population geography refer to the definition of the subject area, and how to increase the influence of the geographical component within population studies. It is suggested that the answer to both questions lies in greater specificity with concentration on key topics, greater use of demographic techniques, and the development of an applied or problem-oriented population geography. Finally, three particular areas in which geographers can make a significant contribution are identified. First, the development of a comprehensive theory of migration. Second, study of spatial variations in demographic structure between cultural regions and their historical development, third, further development of geographical data bases and application of forecasting methods to contemporary issues.

The national census generally represents the most comprehensive and up to date set of information available on the demographic, economic and social characteristics of a population. As such it is a prime data source for researchers. In Chapter 2 John Dewdney and David Rhind seek to introduce potential users to the characteristics of these data sets by describing the nature of the population censuses in Britain and the United States. The British census is one of the longest series of decennial population data in the world. The origin and development of the modern census structure is discussed and the major innovations (e.g. adoption of the household as the basic unit of enumeration) and difficulties (e.g. inter-censal boundary changes) identified. Issues related to the collection and processing of the data are then addressed including, for example, the legal basis of the census, confidentiality constraints, the enumeration district base and the role of the enumerator, census layout, and the means of processing the 1000 million pieces of information on nearly 20 million household schedules. The variety of forms of output covering different geographic scales is described. These range from published volumes (including county, regional, new town, and national reports together with key statistics for local authority and urban areas) to the unpublished small area statistics which contain almost all classifications and cross-tabulations likely to be of relevance to researchers. Analysis of this data has been facilitated by computer packages such as SASPAC. A similar organisational framework is employed to examine the properties and potential of the United States' census. The rapid growth of information contained in the census required the introduction of administrative and technical innovations in methods of collection and processing. These have included the introduction of FOSDIC (film optical sensing device for input to computers), the use of sampling techniques in data collection, and the distribution and collection of forms by public mail. A detailed examination of the structure of the 1980 census is presented, and the problems of its prosecution discussed. Particular attention is afforded to the spatial dimension of the US census and the function of maps, geographic base files/dual independent map encoding, and state master reference files explained. These three elements offer the potential for automated tabulation of results for an almost infinite variety of geographical bases (e.g. SMSAs, school districts, census tracts, enumeration districts and blocks). The content of the US census is considered with perhaps the greatest contrast with Britain being the inclusion of direct questions on income, property value/ rent, and cost of utilities and fuel. For researchers a significant difference is in the cost of obtaining data, with the US census regarded as public domain. Finally, future developments are identified and major contrasts and comparisons drawn between Britain and the United States.

The importance of fertility studies in both academic (e.g. to explain population structures) and practical terms (e.g. predicting demand for services) has not been reflected in the volume of work devoted to this aspect of population geography. In Chapter 3 John Coward seeks to redress this imbalance by examining fertility patterns in the modern world at a variety of geographic scales. At the global level fertility patterns are characterised by significant variations between regions, and in the variables influencing such distributions. Factors suggested to explain the recent incidence of low fertility rates include the diffusion of "modernisation" to some Third World countries, with a key variable being the changing position of women in society, as reflected in better levels of education, increased status and security, and higher proportions of economically active females. Other causal factors are the effects of family planning schemes, cultural and religious beliefs, and government policy and attitudes. Attention is then focussed on the intra-national scale and on an analysis of recent fertility patterns within the British Isles. Significant themes identified for England and Wales are the decline in <u>regional</u> fertility differentials over this century, but the continued existence of fertility differences at the local level. Such findings clearly underline the importance of scale of analysis in geographic investigations. The principal factors contributing to the regional patterns were identified as a decline in social class-based differentials in fertility, and the regional convergence of key variables influencing fertility (such as female participation in the labour force, and the spread of new cultural values and norms). Analysis of the pronounced variations in fertility at the district level examined the contribution of each of the three constituents of fertility (marital fertility, proportions married, and extra-marital fertility) to the observed patterns. Overall it is concluded that while certain demographic and socio-economic factors contribute to variations in age-standardised fertility the level of explanation is far from complete and continued search for additional variables is essential. Similar analyses are presented for the other countries of Britain, and the Republic of Ireland. In the latter, analysis of data for the period 1971-1981 at the county level identified the dominant trend as an extremely high (25%) decline in marital fertility. Possible explanatory factors include a decline in agriculture, the growth of an urban-based middle-class, return migration from Britain, increased employment opportunities for married women, and the increasing strength of feminism and rising secularism among the young. Further progress in understanding fertility trends requires continued analysis of spatial fertility patterns both to identify additional fertility-related variables, and to assess the role of general models of fertility change. Methodological questions of particular importance concern the nature and availability of data and the influence of scale on empirical findings.

In Chapter 4 Peter Curson illustrates the potentially significant role of mortality studies within a modern process-oriented population geography. It is observed that while medical geography has contributed to the spatial analysis of mortality less consideration has been given to the question of differential mortality arising from variations in social class, occupation, ethnicity, race, age and sex. Examination of the social inequality of death supports the existence of an inverse relationship between mortality and social class, (which is itself a summary measure of a range of factors including living and working conditions, nutrition levels, maternal status and access to health care facilities). Significantly, evidence from England and Wales based on standardised mortality ratios for different social groups suggests that the class gradient has steepened over the last half century, a fact which clearly questions the distributive impact of the health care improvements of the last eighty years. Such class-based disparities are of even greater magnitude in the Third World. Significant differences in mortality patterns are also found between racial and ethnic minorities and the majority population in Developed Countries with, in some cases, the incidence of childhood mortality comparable to rates in the Third World states. At a general level a clear distinction in major cause of death exists between western degenerative diseases (such as heart disease) and the parasitic and infectious diseases (e.g. gastroenteritis and tuberculosis) which dominate mortality patterns in the Developing World. The spatial patterning of mortality has attracted most geographical attention and examples of this research are provided at the international, regional and intra-urban scales. While the mapping of mortality distributions can describe particular environments of risk, indicate possible lines of epidemiologic enquiry, and suggest modifications to health care programmes causative explanation represents a more elusive goal. Some progress has been made in identifying risk factors in the natural and social environments which contribute to premature death. These generally include climate, air and water pollution and housing conditions, while in a Third World context, the role of environmental factors, land use practices and the effects of the development process have all been related to disease hazard. The central problem, however, remains that of establishing a causal link, (as opposed to a statistical relationship), and it may be that "the scientific demonstration of causality in geographic research remains an illusory and unobtainable goal". It is suggested that longitudinal and cohort studies and reconstruction of personal histories may offer greater promise than current residence-based studies of mortality. It is also necessary to incorporate a greater range of possible factors (e.g. biochemical or physiolergic influences) into the analysis. It is concluded that more sophisticated methodologies which recognise the temporal as well as spatial dimensions of disease, and investigations at more detailed levels, concentrating on particular population groups are required if we are to approach the objective of understanding the processes which lead to spatial variations in mortality.

All governments enact policies which, either directly or indirectly, affect the natural growth and distribution of their populations. Population policies can seek to achieve a number of objectives, ranging from urban decongestion to national development. The strategies selected reflect the views of government and internal pressure groups as well as those of international agencies and researchers. In Chapter 5 Ian Thomas examines the relationship between population research and government policy, and explicitly considers the contribution of a geographical approach to the analysis of the components of population policy. It is argued that while other disciplines also examine the factors associated with population dynamics geography has a particular responsibility to explain regional variations in the operation and outcome of these forces. Attention is then focussed on the nature and origin of population policy in the United Kingdom where concern in the inter-war period over the uneven regional incidence of unemployment and low levels of fertility was replaced, in the post-war era, by policies with anti-growth tendencies (e.g. family planning programmes and immigration controls). In assessing the contribution of geographical studies to the formulation of population policy three areas of research are considered. The first refers to policies designed to affect the rate of population growth and in this area the geographical input is significantly less than that of the other social sciences. Potential areas for geographic analysis include the diffusion and utilisation of family planning, the influence of spatial-cultural variations in the acceptibility of policies, regional and social differentials in mortality, and the policy implications of concentrations of elderly or ethnic populations. Policies designed to affect population distribution and redistribution have been more directly influenced by geographic investigation in terms of both migration theory and empirical studies of migration motives, migrant characteristics, and the impact on origins and destinations. This is illustrated with reference to geographical research into African populations. The third main research area considered comprises geographic studies which affect policy indirectly. These include regional studies of population as well as thematic investigations of housing and transport issues and methodological advances in data collection and analysis. It is observed that, over recent decades, governments in both the Developed and Developing World have displayed increasing interest in the study of population dynamics in order to acquire information on which to base national planning priorities. There is also concern over inequalities at regional and urban levels, and a general interest in the relationship between population distribution and social and economic development. Such trends emphasise the relevance and potential of a geographical approach to population policy.

Migration is a human activity which occurs at all geographic scales from local to international, and the analysis of migration flows represents a major area of research in contemporary population geography. In Chapter 6 John Salt examines the contribution of the spatial approach to the analysis of international migration, considering both voluntary labour and involuntary refugee movements. Problems relating to data and definitional limitations and the absence of an empirically-verified universally applicable model of international migration are considered. An analytical framework is proposed which recognises that (a) geographical patterns of international migration occur within a systems framework and are related to complex networks of interaction, and (b) combinations of global and local forces produce spatial migration networks which differ in scale and behaviour. The increased explanatory power gained by the move from macro-scale models based on social physics towards process-oriented micro analytic perspectives on migrant behaviour is also underlined. The nature of political barriers to international migration is then discussed before the characteristics of a series of migration networks is examined. It is emphasised that in order to understand the complexity of international systems, knowledge of aggregate labour market conditions must be complemented by information on the socio-economic and demographic factors which underlie population movements. The phenomenon of return migration has attracted considerable interest in recent years. To date, however, research has concentrated upon the economic dimension, and more detailed investigation of the heterogeneity and differing motivations of migrants is required. Specific attention is then afforded to the migration of high-level manpower. It is suggested that variations in the type of personnel involved and changes in the nature of flows over time reflect government strategic requirements and policies, the nature of economic linkages between countries, and the labour policies of employers, particularly multi-national companies. In contrast to labour migration the enforced movement of refugees is an under-researched component of international migration, with most work to date focussing on the consequences rather than on achieving theoretical understanding of the causal forces. Attempts which have been made to conceptualise the determinants of refugee flows are considered, ranging from typological structures based on the characteristics of refugees to analyses of the political forces generating movement. A model of the global refugee system is presented as a potentially useful explanatory framework for the analysis of refugee flows. It is emphasised that while political factors generally underlie refugee movements the varied nature of the resulting flows defines a clear role for geographical analysis. The spatial approach to the study of international migration is characterised as both empirical (ie. data dependant) and pragmatic (ie. concerned with real world problems). It is concluded that the major contribution of the geographical approach lies in its ability to identify variations within and between migration networks, which both highlights the danger of overgeneralisation and helps define the limits of any general theory of international migration.

In Chapter 7 Joseph Gugler examines the phenomenon of internal migration in Third World countries where the predominant characteristic is population growth in both urban and rural areas. Two major migratory patterns are evident; rural - rural movements (e.g. in response to pressure on the land resource, or due to natural disasters), and rural-urban migration. The latter is the subject of particular attention, given the social, economic and political problems posed, and the crucial role of this migration process in the global urban transition. The effects of the incorporation of traditional economies into the world capitalist system are then discussed. These include the accelerated growth of population and an increased significance of the urban-rural gap in living standards. The nature and extent of these differences, measurement problems, and the necessity of disaggregating data to consider the position of individual migrants are discussed. It is suggested that urban-bias on the part of government decision-makers is a major causal factor with rural peasants, effectively deprived of political and economic representation, resorting to migration in an attempt to improve their quality of life. The fact that the fundamental factor underlying rural-urban migration is economic leads to examination of the urban labour market. It is observed that whereas in the colonial period employers experienced high rates of labour turnover and absenteeism, as economies were incorporated into the world system labour shortages were replaced by oversupply, unemployment and underemployment. The dynamics of this radical change are explored. Variation in the structure of urban labour markets also underlies two interpretations of migrating behaviour. The first postulates a random process of labour recruitment characterised as an urban job lottery, in which relatively rapid turnover and high minimum wages made even an extended job search a worthwhile strategy. This however appears to have been an exceptional model, typified by tropical Africa in the post-colonial era. More commonly, labour recruitment is structured on the basis of a set of segmented markets with access to job opportunities dependent upon the migrants' education, sex, age, training and patronage. It is emphasised that the flow of migrants into cities represents only a fraction of the potential migrants in the rural reservoirs. Three principal patterns of migration are identified, return migration, circular migration, and permanent settlement, - and the operation, advantages and disadvantages of each strategy are examined. Finally, consideration is given to government attempts to reduce the rural-urban flows. It is observed that in practice internal migration controls are difficult to enforce and are of limited effectiveness. It is argued that policies which seek to deny the benefits of urban life to disadvantaged rural dwellers are socially regressive, and that what is required is a positive policy aimed at the development of rural areas. This would require a significant re-orientation of resources (e.g. via the tax system, agricultural pricing or direct investment) and run counter to the interests of urban-based power groups. This suggests that such a strategy is unlikely to be widely adopted and that rural-urban migration will continue to be a feature of the population geography of the Third World for the forseeable future.

For most of this century the dominant direction of population movement in western capitalist societies was from rural to urban areas. Since the early 1970s, however, a reversal of this trend has become apparent and the phenomenon of counterurbanisation has become a major component of contemporary population dynamics. In Chapter 8 Tony Fielding examines this process in a representative array of countries and then considers the various explanations for the emergence of such a pattern. Attention is given first to the evidence for the "population turnaround" in USA where dominant elements in the post-1970 population redistribution included, (a) a negative relationship between population growth and urban size, (b) a positive relationship between population growth and distance from the city centre, and (c) a net migration gain by non-metropolitan areas at the expense of metropolitan areas. The existence of similar trends in other countries is revealed by detailed consideration of Australia, Japan and France. In Australia, while the turnaround was more than a simple reversal of the previous pattern of urbanisation (with, for example, population loss experienced by remoter rural regions as well as large urban centres), common features included a net internal migration loss from the major cities and the emergence of a "sunbelt" phenomenon in Queensland. Similarly in highly-urbanised Japan, and in France, a reversal of migration flows in favour of rural areas became apparent around 1970. Attention then turns to critical analysis of three major sets of arguments to explain counterurbanisation. The first suggestion that it is a statistical artefact arising from the underbounding of de facto urban areas is rejected since it is clear that the process involves more than suburbanisation. A second paradigm suggests that place preferences can best explain the changing patterns of net migration. It is concluded, however, that since migration decisions are normally constrained by income-related factors this argument has at best a supporting role and has greatest relevance for retirement migration and tourist employment. The third model suggests that counterurbanisation results from world economic conditions (in particular the recession and energy crisis) which improved the position of rural regions relative to metropolitan areas. This argument, based on parallels with the 1930s, is undermined by significant temporal differences (such as, the current availability of social welfare programmes) and evidence which suggests that the population turnaround commenced prior to the economic crises of 1973. The analysis concludes by identifying the explanatory power of an approach based on the geography of production. Utilising the concept of the spatial division of labour this suggests that counterurbanisation may be due to a shift from regional sectoral specialisation (in which each place concentrated on all tasks involved in producing a good) to a new spatial division of labour in which places are differentiated with respect to the part they play in the production process. This suggested that the dominance of RSS up to 1960 produced urbanisation whereas since the 1970s the NSDL has allowed some forms of employment (e.g. routine production) and employees (e.g. managers) to decentralise. Further empirical evidence is required to support the RSS-NSDL transition model but preliminary indications are that it offers a potentially useful analytical framework.

In Chapter 9 Martin Cadwallader examines recent research into inter-regional migration and intra-urban residential mobility. The spatial pattern of regional migration flows and the relative importance of push and pull factors have been the subject of investigation ever since Ravenstein formulated his "laws" of migration. While the importance of distance and direction has been demonstrated it may be questioned whether their individual effects can be isolated from those of other factors. Attempts to identify the relative importance of different determinants of migration using the gravity model and single equation models have indicated the relevance of unemployment rates, income levels, age and education differences, and previous migration patterns as well as environmental factors. Several statistical (e.g. multicollinearity) and theoretical problems (e.g. the two-way relationship between migration and the explanatory variables) are identified and possible solutions suggested. These include field theory, cannonical correlation analysis, and the use of structural equation models which specify both the linkages between dependent and independent variables and also the interrelationships among the initial causal variables. The contribution of dynamic models to the study of temporal change in migration patterns (e.g. via Markov chain analysis or time-series techniques such as moving average and autoregressive models) is also discussed. Attention then turns to the behavioural approach to migration, centring on the investment in human capital theory and on the concepts of place utility and residential preference. In addition to the links between preferences and migration behaviour the nature of preferences has been explored and the utility of several procedures (such as conjoint measurement techniques and multi-dimensional scaling) is illustrated. It is suggested that further progress in understanding inter-regional migration may be gained by combining macro and micro-scale approaches. Several important research questions are identified including (a) the study of temporal changes in migration patterns, (b) incorporation of individual and institutional constraints into migration models, and (c) greater attention to the consequences as well as the causes of migration. The discussion then turns to consider residential movement within cities. Spatial variations in mobility rates are identified and possible explanatory factors, (related to demographic, social, economic and housing characteristics of the urban environment) are examined. As at the regional level, aggregate analysis of the role of distance and direction in intra-urban movement is complemented by consideration of behavioural models which focus on the individual migrant. It is observed that although they are interrelated, the major components of the migration process can be examined singly in an attempt to illuminate the underlying mechanisms. The decision to move, for example, has been investigated using concepts such as residential stress. Study of the search process has considered information sources, optimum stopping rules and the location of search activity; while attempts have been made to model residential choice processes by means of log linear models. In most formulations to date, however, insufficient attention has been given to the constraints on individual migration. It is suggested that further investigation of the effect of housing supply constraints on migration behaviour could increase the predictive ability of existing models and enhance their utility in terms of public policy formulation.

The explanation and prediction of human behaviour in space is the general objective to which quantitative population geographers seek to make a contribution. Two specific areas in which population modellers have shown interest are the behaviour of populations in the past and how population may develop in the the future. In Chapter 10 Philip Rees, concentrating on predictive formulations, explains the principal conceptual and methodological issues underlying population modelling. Four fundamental concepts are examined. First, the concept of the ideal data set and life history frameworks, second, the states of the system, third, the type of transfer data and model type, and fourth the age-time plan used in observing or measuring demographic variables. The properties of an ideal data set embodying the life histories of the population are detailed, and the potential utility of the OPCS longitudinal study noted. The chief importance of such a conceptual framework is that it enables researchers to evaluate their data against an ideal and so recognise any deficiencies or biasses. The second key concept refers to the ways in which a population is classified and to the movement of individuals between such classes over their lifetime. Several examples of the kind of state to state transfers that have commonly been built into population models are provided (e.g. between age groups, marital condition, labour force status, and geographic areas) and the problems of operationalising these are considered. The difficulty of incorporating several dimensions into a model is examined. Thirdly, the way in which different kinds of transfer data affect the calculation of multi state transition probabilities and influence the interpretation of model outputs is discussed. The basic distinction is drawn between transfers as transitions, moves, and last migrations. It is emphasised that in developing a multi state model of a population it is first necessary to ascertain the type of transfer data available and then adopt the appropriate model structure. The use of the method of demographic accounts in estimating multi state mobility is then illustrated. The final set of concepts considered concerns the age-time plan (ATP) employed in observing or measuring demographic variables. The nature of four ATPs is explained with the aid of Lexis diagrams, and the benefits and problems of employing each are identified. Attention then turns to two fundamental models in population geography - the cohort survival and the life table models - and how these can be linked to the concepts outlined. A general formulation for multi-state projection models is then presented and critically assessed. Some problems in implementing the projective models devised are identified and possible solutions suggested. To be successful population modelling must be based on sound information on the nature of the system being modelled, on the type of population states necessary for the problem, and on the type of transfer data available for the system under investigation.

Chapter One

THEORY AND METHODOLOGY IN POPULATION GEOGRAPHY

R. I. Woods

Population geography has no one theory, methodology or, for that matter, definition. It is a widely recognised sub-field of geography as a teaching subject and work on population matters is prominent in the research discipline that is geography. But the distinctions between geographies of population, where population is merely one aspect of a complete human geography; spatial perspectives in population studies or demography, and population geography as a separate entity in its own right are exceptionally blurred. Some of these difficulties relate to the inherent problems faced in defining geography, but others stem from the nature of population studies as a diffuse multi-disciplinary specialism.

In order to clarify what are the important elements in this chaotic state of affairs we must begin by examining the origins of competing definitions of population geography. This will provide an introduction to what will be the main concern of this chapter: a discussion of the theoretical and methodological developments in both the geographer's approach to population and, since it may be that population geography is also being done by non-geographers, population studies in general.

DEFINITIONS

Glenn T Trewartha's presidential address to the Association of American Geographers in 1953 provides a convenient starting point for our consideration of the various definitions of population geography. Trewartha (1953, see also Kosinski, 1984) was concerned with the need for geographers to treat population matters as a principal sub-discipline (with cultural and physical geography) of a teaching and research subject that was essentially holistic in form. But apart from bemoaning the neglect of population he proposed 'a system for population geography' which even thirty years on, is still well worth considering. 'The geographer's goal in any or all analyses of population is an understanding of the regional differences in the earth's covering of people. Just as area of differentiation is the theme of geography in general, so it is of population geography in particular' (Trewartha, 1953, 87). Population geography is the 'area analysis of population', and so Trewartha's 'system' lists 'the kinds of population features to be observed and compared in different areas'. A diagramatic summary of Trewartha's (1953, 88-87) 'system' is illustrated in Figure 1.1. The objective, shown in the top-centre box, is to understand the area differentiation of population, but this is accomplished via at least three routes: historical, numerical and qualitative. The second box in Figure 1.1 seems the most important since it covers mortality and fertility variations (2b): the distribution of population, settlement size and urbanisation (2d): migration at various scales (2e): but is also linked with notions of resources and carrying capacities (2c). The third box relates to area differentiation in the qualities of population; some of which are physical (3a) and some socio-economic (3b). The point needs to be emphasised that Trewartha's 'system' would regard area differentiation in, for example, levels of economic development as part of population geography. This is entirely consistent since the other two thirds of the geography triangle are physical and cultural, neither of which would presumably deal with economic development as a central issue.

Although Trewartha argued the case for population as the 'pivotal element in geography' he was also aware that, 'there is bound to be lack of agreement on the full content of the field of population geography and admittedly there is no one way or best way of ordering and arranging the topics to be included. Still, in most disciplines and branches of disciplines, there is a core of content on which there is reasonable agreement, even though the full content and its arrangement may bear the stamp of individual authorship' (Trewartha, 1953, 87). Despite changes in geography as a teaching subject and research discipline many authors have used Trewartha's 'system' over the last thirty years. John I. Clarke's (Clarke, 1965 and 1972), approach has been particularly influential. 'Population geography... is concerned with demonstrating how spatial variations in the distribution, composition, migrations and growth of populations are related to spatial variations in the nature of places', and thus 'population geographers endeavour to unravel the complex inter-relationships between physical and human environments on the one hand, and population on the other. The explanation and analysis of these inter-relationships is the real substance of population geography' (Clarke 1972, 2; see also Clarke, 1984). More recently, Robin J. Pryor (1984) begins his review of methodological problems in population geography with: 'It is assumed here that population geography deals with the analysis and explanation of interrelationships between

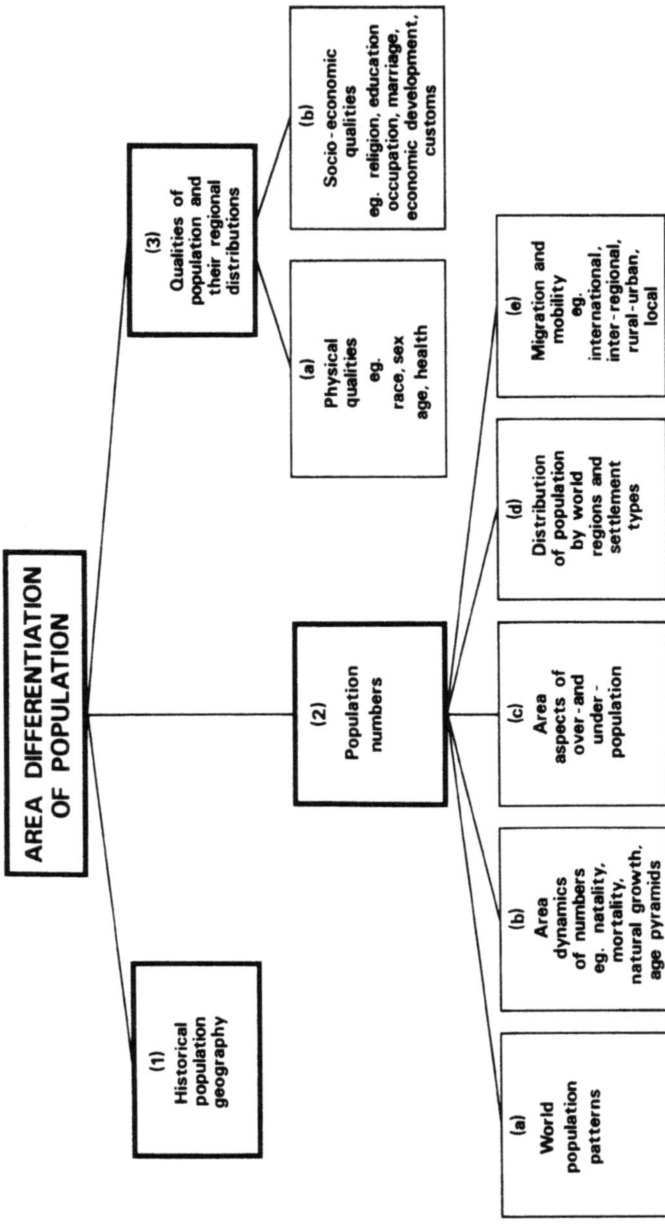

Fig. 1.1 A diagramatic summary of G.T. Trewartha's 'system for population geography'. (Source: based on Trewartha, 1953, 88–89)

population phenomena and the geographic character of places as both vary through time and space. Population phenomena include the dynamics of population distribution, urban/rural location, density and growth (or decline); mortality, fertility, and migration; and structural characteristics including age, sex, ethnicity, marital status, economic composition, nationality, and religion' (Pryor, 1984, 25). Here we have a workable and unified definition that links Trewartha's 'system' with Clarke's concern with spatial variations in populations and places.

This broad definition is also reflected in Daniel Noin's Geographie de la Population (1979) and Jürgen Böhr's Bevölkerungsgeographie (1983). Both authors are concerned with the spatial distribution of population, with the components of its growth and with the characteristics of populations. Noin's (1984) survey of the contents of population geography textbooks published during the last twenty years shows quite clearly that the broad definition has been the one most widely adopted and that its chief distinguishing characteristic is a close adherence to Trewartha's 'system' but especially boxes 2d, 3a and 3b. Other definitions and approaches have been proposed, however.

The narrow definition is to be found in Hoods (1979, 1982) and Jones (1981). Their approach, both stated and implied, is largely concerned with material related to boxes 2b and 2e in Figure 1.1. That is variations in population dynamics approached via an analysis of the mortality, fertility and migration components at various scales, in the case of Jones (1981), and more generally the study of population from the spatial perspective. The narrow definition deliberately neglects population distribution and redistribution (2d) and the 'qualities of populations' (3), except in the sense that the latter affects the components. This attempt to narrow the field of view and to develop a demographic geography or spatial demography has not gone unchallenged by the advocates of the broad definition (see, for example, Clarke, 1980, 1984) who, as we have seen, still adhere to Trewartha's 'system' of Figure 1.1. They would answer Clarke's (1980, 368) question, 'How much more demographic should population geography become?' in a way that would accept the techniques of analysis developed by demographers, but would reject the wider methodology and objectives. Newman and Matzke (1984) seem to suggest a sensible compromise. For them 'population geography is a relatively open field of inquiry. It does have a recognisable core, but there is considerable room for many issues that relate to people and their well-being' (Newman and Matzke, 1984, 6). However, the core comprises the demographic variables, population change and distribution; while beyond may lie 'social and economic indicators' (e.g. language, ethnicity, religion, occupation); residential characteristics' (rural and urban); and 'population in its broader human context' (e.g. resources, politics, policy).

It might be objected that an extended debate over the definition and content of population geography is bound to be arid at best, destructive at worst, that population geography should be what geographers active in teaching and research on population do. There is no need for either a broad or narrow definition where inter-disciplinary boundaries are being crossed with increasing ease and by growing numbers. All this may be so, but one is left to ponder a riddle set by Clarke (1977, 136), 'Why has the academic influence of population geographers been less significant than their numbers would suggest?' 'Academic influence' will be taken to stem from the theoretical insights provided by a discipline or sub-discipline together with the empirical knowledge, but perhaps especially the useful knowledge, accumulated and the ability to predict successfully. Using these criteria it may be appreciated why geography as a research discipline lacks academic influence; why within the discipline the influence of population geography and population geographers has been relatively insignificant; and why the same could be said of geography's influence on demography or population studies. As Jones (1981, vi) has remarked, 'Population geographers have not figured among the stormtroopers of methodological transformation in geography in recent years...'

My answer to Clarke's riddle is to argue that because of the broad definition population geographers have spread themselves too thinly over too large a field; that generally they have not mastered the techniques of demography as Trewartha (1953) advised them to; and that they have been unadventurous in their willingness to develop new concepts for understanding and prediction. Of course there have been exceptions and there are signs of change, but most are very recent. I shall consider the theory and method aspects of this argument in the next section, and only two examples of the exceptions must suffice at this stage. Regional population forecasting, once described as having remained peripheral to population geography (Clarke, 1980, 389), has shown itself to have great vitality and practical value (see Hoods and Rees, 1986). Similarly the ability of population geographers to assemble and manipulate large data sets of useful knowledge in ways that are convenient to other researchers, planners and decision makers can only enhance their influence (see, for example, Rhind, 1983).

I shall return to this question of academic influence later, but at this stage it is necessary to repeat that there is no one agreed definition of population geography; that the so-called broad and narrow approaches are not mutually exclusive, rather they represent differences of emphasis, and that the persistence of uncertainty may prove a source of weakness. We must now turn to consider the main theme of this chapter, theoretical and methodological developments, but the issue of definitional diversity will remain in the background.

THE NEED FOR THEORY

All the advantages of working with a theoretical framework need not be spelt out here, but it may be worth emphasising four particular points which are helpful in population geography. First, the use of theoretical statements, that is statements of prospective association, help to focus research which might otherwise become diffuse, aimless or over-empirical. Second, the use of a broad theoretical framework permits inter-disciplinary comparison to be made more forcibly. Third, the consciously interrogative style of theory construction and destruction aids the definition of priorities and creates a sense of progress when a succession of partial solutions are forthcoming. Fourth, the adoption of a theoretical framework makes explicit the need to tackle issues of purpose, meaning, understanding, explanation and interpretation which might otherwise remain at a subconscious level.

An effect of recognising these points will be to push population geographers closer to the mainstream of methodological debate that has infused human geography in the 1980s. Figure 1.2 indicates some of the issues that will need to be tackled. The upper left pane of the window shows examples of the levels of theory available in general and emphasises both the need to agree upon the purpose of theory construction and the criteria to be used in theory validation (how theories are to be judged). The lower left pane lists five distinctive methodologies which are currently in use amongst geographers, but it does so in a way that suggests a sequential order from empiricism to structuralism which is at least partially in step with recent developments in human geography (Johnston, 1983). The two panes on the right have been left to be completed by the reader, but the rest of this and subsequent chapters will provide discussion of a to e and examples of 1 to 5.

Purpose

Consideration of purpose not only returns us to definitional issues introduced above, but it also makes us aware of the differences in levels of understanding that are being sought. For example, there is a clear distinction between studies that seek to model or forecast migration flows using a black box technique and those that attempt an interpretation of human behaviour and decision making with respect to mobility. The former used theory in a modelling sense; they are concerned

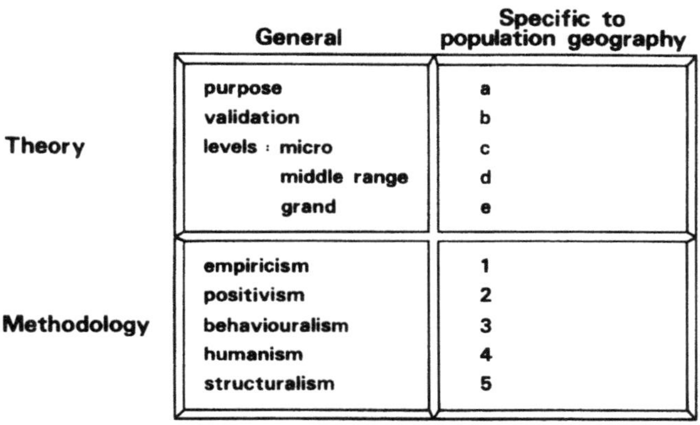

Fig. 1.2 A theoretical window.

with the relationships between output and input, with responses to parameter changes and with the need to predict future patterns and distributions. The latter have more intangible objectives. They deal with notions like place utility, satisfaction, environmental stress, stage in the life cycle, bounded and selective rationality and constrained action. These studies seek a broader understanding of why individuals or groups migrate and where they move to, but there are no agreed criteria on what understanding should mean. While some researchers deal with the correlates of flow patterns (Greenwood, 1981) others are concerned with the characteristics of movers compared with stayers (Speare, 1971, White and Hoods, 1980) or the psychology of decision making and implementation (Wolpert, 1964, 1965, 1966; Lieber, 1978; Desbarats, 1983). All contribute to the common stock of information on migration, all use theory to some extent, but even though the ultimate objective can be stated, methods are as diverse as the individual researchers using them.

As far as the use of theory by geographers studying migration is concerned while the general purpose is clear there is little evidence that the diversity of short-term objectives and methods has been explicitly recognised or that the variable status of conclusions has been assessed in ways that reflect such diversity.

Validation

How are theories to be assessed? Can they be proved or merely temporarily accepted while evidence for conclusive rejection is lacking? These are important questions which require careful consideration when theoretical frameworks are used in population geography, yet they are often either ignored or the alternative forms of answers are not recognised.

The broad definition of population geography includes settlement geography (2a in Figure 1.1) and thus central place theory. The history of the development and use of Christaller's theory provides an interesting case study of the rise and fall of a very particular form of theoretical framework. Classical central place theory is a normative-deductive theory based on a number of specific environmental and behavioural assumptions. Once the stage has been set and the cast selected the play unfolds in a highly ordered fashion, the final outcome is clear and the progress to it inexorable. Interest in Christaller and Losch's versions of the theory was rekindled in the 1950s and 1960s amongst geographers who sought to move beyond the description and classification of settlement types who required for their purpose a rigorously developed theory of the location and function of settlements. The first inclination of these geographers was to 'test' the theory. That is to derive from central place theory hypotheses which could in turn yield empirically verifiable predictions. Surprisingly many of the early tests revealed that there was rather more regularity in settlement patterns than one might have expected a priori and that certain aspects of urban functional hierachies could be identified (Berry, 1967; Berry and Horton, 1970). Where there was a poor correspondence between observed and expected, geographers were likely to relax the assumptions on which the theory was based. Since even in Iowa and East Anglia the isotropic plains were bounded and undulating how could one expect perfect hexagons to emerge? The rationality of man as homo economicus also came into question. Was motorised man indifferent to the cost effects of relatively short increases in journey to purchase and thus capable of ignoring the nearest retail outlet for a desired good? Even if one accepts an affirmative answer it must be recognised that the economic ordering of space is not entirely chaotic and thus that settlement patterns may prove to be non-random.

All of this presupposes that classical central place theory is expressed in a form that is capable of empirical verification. But this may prove a dubious position to take. In its purest form central place theory is beyond verification, rather it provides a theoretical framework of the ideal type form. It is constructed in such a way that the comparison of observed and expected will mainly serve to highlight irregularities in the observed pattern since the assumptions on which the theory is based are highly idealised. The theory itself may only be assessed in terms of its internal logic, that is the manner in which conclusions can be consistently derived from initial assumptions. Once the theory has passed this test, its main function is to reveal those non-economic aspects, such as environmental variation and unequal resource allocation, that distort the purely economically determined form of settlement pattern generated by central place theory (see Webber, 1972, 88-116; Leibenstein, 1976).

This example serves to raise some interesting general points regarding the form of validation to be used in population geography. It suggests the need for flexibility and the value of avoiding starkly defined categories - true-false, accepted-rejected - and above all it refers us back to the question of purpose, what is our objective in constructing a theoretical framework?

Levels: Micro-theory

The matter of purpose also lead us on to consider the levels at which theories can be constructed. Merton (1967, 39) has distinguished three distinct categories which he labels micro-theory, middle range theory and grand theory. Each of these levels relates to a particular way of constructing knowledge. For example, and to oversimplify, micro-theory deals with the behaviour of the essential building blocks of a society or economy in their individual forms. Here one seeks to understand the actions of individual persons, or families or households, how they think, perceive their relative positions, act to change and react to their circumstances. Middle range theories are concerned with the characteristics of aggregates, groups of people defined in spatial, occupational, ethnic etc categories. This form of theory deals explicitly with dependent and independent variables, with the multivariate association between a number of the latter and one of the former. Grand theory is rather more difficult to define precisely, but its sphere is typified by long-term social, economic and demographic change. For population geographers demographic transition theory provides perhaps the most obvious examples (see Skinner, 1985, for a general discussion).

Interest in micro-scale theory amongst population geographers has related mainly to their concern with migration studies, their desire not only to describe the characteristics of migrants, origins, destinations and flows, but also to understand how and why individuals decide to move (White, 1981). But these ideas may also be appropriate when dealing with other demographic events (when and whom to marry, how many children to have and when, whether to get divorced). Let us consider the case of marriage, a rather neglected area of population geography.

Cross-cultural definitions of marriage encounter certain problems for the significance and meaning of marriage varies in important respects between societies. The most important aspect relates to the legal rights of children born to parents who are married. Marriage thus becomes an institution which legitimises reproduction, involves a more or less permanent sexual union (implying that both partners must be sexually mature) and the onset of which is signified by a public ceremony. In many African and Asian societies marriages are arranged between teenage brides and grooms by parents often with the assistance of third-party marriage brokers. After the marriage ceremony the bride joins the household of the groom's parents where she is in a sense 'on probation' until the birth of her first child whereupon she becomes a fully accepted member of the parental family and household. In these circumstances the rules and conventions of marriage are clearly determined, everyone knows what is expected of them and deviation is exceptional. In Westernised societies marriage has the same reproductive and legitimising functions, but the arrangement of marriage is far more haphazard. Here there is a balance between the choices of individuals, and the constraints imposed by parents and society in general. The social norms involved relate to sanctions against non-marrying couples and the use of a family orientated ideal as a means of giving meaning to human existence. Individual choice, on the other hand, influences the age at which marriage occurs, the partner selected and the duration of the marriage. For the demographer it is the age at marriage and the proportion of the population, especially female, married in any one age group that is of most interest since these elements may have important effects on fertility patterns particularly where family limitation is not in use (Woods and Hinde, 1985). But the population geographer may also be concerned with contact patterns and information fields (Perry, 1969; Morrill and Pitts, 1967).

Any micro-theory of marriage will need to contain two elements: search and selection. Let us suppose that once an individual becomes eligible (at 18 years perhaps) he or she will enter the search category which contains passive and active searchers (the proportion of the latter increasing with age). The search will itself be affected by opportunities, such as the sex ratio amongst eligibles which is demographically determined, and by the structure of day-to-day contact patterns which are largely conditioned by a searcher's residential, employment and recreational characteristics. Mutual selection, on the other hand, will be influenced by perceived compatibility on a number of levels (physical, educational, interests, family background, for example). The decision to marry will thus represent the culmination of a process of search and selection which at every stage biases the outcome towards the pairing of mirror images. But marriage also represents a calculated decision to desist from further search, to capitalise on present advantage rather than seeking for additional gain in the future (see Becker, 1981).

Viewed in this way marriage and migration will be seen as manifestations of related phenomena which can be understood within the context of constrained choice. The actors are rational to a degree, they base their decisions on specific criteria which although they may not be articulated are revealed by preference patterns, but they must deal with both a limited appreciation of the present and an inadequate knowledge of the future. The search is biased and the selection faulty.

Levels: Middle-range Theory

It has already been noted that middle-range theory tends to deal in aggregates, with the characteristics of groups of people variously combined. Richard A Easterlin's (1969, 1971, 1978; Easterlin, Pollak and Wachter, 1980) attempt to reconcile economic and sociological theories of fertility into one composite theory provides an interesting illustration of what may be attempted with this level of conceptualisation. Easterlin draws from economics the notion that the number of children born may be a matter of parents balancing costs and returns; that when the supply of children exceeds the demand then the use of means to reduce the number of additional unwanted births will be affected by the costs of family limitation. The supply of children is taken to be biologically determined, but because Easterlin (1978) mainly deals with surviving children his formulation is affected by changes in infant and child mortality. The demand for children will be related to social norms (children having a value for their own sake), but also the rising costs of rearing 'high quality' children (prolonged education etc) and their diminishing value as contributors to the family economy. Easterlin argues that in pre-modern societies there will be a high demand for children and that it will exceed supply, but that with modernisation supply will rise and demand will fall. Eventually supply will meet demand as effective means of birth control both become cheaper and are used to avoid unwanted births. Easterlin (1971) used a similar theoretical framework to tackle the question, does human fertility adjust to the environment? Here the environment is divided into three: the frontier, settled rural communities and urban communities. The prediction is that fertility will be highest on the frontier where the demand for children as a source of additional labour will be highest and the means of affecting fertility least accessible, whilst in the urban areas fertility will be at its lowest for there children will not be in such demand and means of birth control are likely to be more available. The settled rural communities are thought to occupy an intermediate position along the continuum. Broadly speaking, Easterlin finds that his predictions are borne out in nineteenth-century mid-west America.

There are two important points here. First, Easterlin draws from an existing, well established theoretical framework principles which he applies to a particular case, aggregate demographic behaviour. Second, the theory so constructed yields empirically testable hypotheses (fertility will be higher/ lower in x when y holds etc) which make the theory falsifiable. These represent two significant aspects of middle-range theories. Middle-range theories will of necessity have to face a number of common problems, however. Most of these relate to validation, but some also occur because groups of people, or areas are used as reference units. For example, the ecological fallacy and the problem of equifinality both delimit formal testing procedures. Despite these often obscured drawbacks middle-range theory represents the most commonly employed level of theoretical device. Additional examples will be discussed below (under 2).

Levels: Grand Theory

Amongst population geographers the most widely used grand theory is undoubtedly that of the demographic transition. In its original form, as conceived by F. W. Notestein (1945), the theory had distinct sections. The first comprised a description of the demographic situation in pre-Second World War Europe and America. The second sketched changes in birth and death rates such that both were initially high, death rates fell while birth rates remained high, and finally, birth rates fell to match death rates at low levels. The third defined three stages in terms of population growth rates: slow, rapid and slow. The fourth and final section proposed an explanatory mechanism (agricultural, industrial and sanitary revolutions) which forced mortality down and ultimately, via the spread of contraception, removed the props to high fertility. In subsequent reformulations Notestein's theory has been enlarged and made more flexible, but it still retains its basis in historical analogy, its three stages and its insistence that demographic change results from cumulative socio-economic advances (often termed modernisation). All forms of transition theory, whether original or revised, deal with long-term change in the entire demographic system which they regard as a consequence of significant changes in the economic technological and political ordering of society. The familiar representation of the theory in model form (time series for birth and death rates) and the inevitability with which successive populations replicate its general shape have given the explanatory section of the theory an immortality which is no longer deserved (see Woods, 1979, 6).

There are three reasons for taking this view. First, demographic transition theory may be said to be too general, it lacks the specificity of a middle-range theory and is thus difficult to evaluate. Second, there is no mechanism within the theory that will generate the sequential movement between stages (3 must follow 2, 2 must follow 1). Third, and perhaps of most importance, the use of historical analogy is likely to provide a faulty logic (what will happen in x will happen in y). It is now obvious that the demography of historical societies was both more varied and more complex than had been previously been thought; that particular cultures were likely to have their own specific demographic regimes; and that demographic experience might not be culturally transferable. For example, it appears that economic development coupled with administrative organisation were mainly responsible for initiating the secular decline of mortality that began in Europe and North America in the late nineteenth century, but in Latin America, Asia and Africa, where mortality has fallen it has done so largely via the introduction of European and American technological experience (especially in terms of medical science, chemotherapy and public health) and has not necessarily been linked with economic growth. In Europe the fall in child then infant mortality tended to reinforce the need for and advantages of family limitation, but the sudden fall in mortality amongst most Third World societies could only lead to rapid population growth since nuptiality and marital fertility were incapable of adjusting so quickly to such dramatic changes in the absence of required social and economic developments. The subsequent fall in marital fertility which has been observed amongst populations in Latin America and certain parts of Asia (especially the island states) relates not only to the effects of urban-industrial growth, but also the various family planning programmes introduced with government support and Western encouragement. In China the fertility transition is being planned independently of modernisation. (There is a voluminous literature on these issues, but Caldwell, 1982, Bulatao and Lee, 1983; Woods, 1982, provide introductions.) One way to begin the development of a grand theory of demographic behaviour would be to argue in terms of the factors that could, in abstract, influence demographic structures. The following provides an example.

A. Factors that will tend to reduce mortality
   1. Advances in medical and drug technology
   2. Economic development
   3. More equitable income distribution
   4. Improved quantity and quality of food supply
   5. Improved quality of water supply
   6. Effective local and national government
   7. Improved education (especially health education)
   8. Political stability

B. Factors that will tend to reduce marital fertility
   1. The availability of effective contraceptives
   2. Urbanisation
   3. Industrialisation
   4. Improved education (especially for women)
   5. Rising social status of women

6. Decline in infant and child mortality
7. Restrictions on the economic value of children
8. Rising living standards

Although many of these factors are associated, change in each one is capable of effecting a negative change in mortality or marital fertility. Demographic transition theory consists of a fixed list of factors which are thought to be universally appropriate. The scheme provided here has a more flexible basis which allows for the substitution of factors. For instance, marital fertility may be reduced as a result of B1 + B4 in the absence of B2 + B3. No one is on its own sufficient or necessary, nor is the exclusion of a factor in a particular case evidence for its irrelevance elsewhere.

The function of grand theory in this instance would be to establish the lists of factors, to specify the associations involved (positive or negative) and to make clear how the demographic variables (mortality and marital fertility above) are themselves interrelated in some total demographic system. It would be for middle range theory to hypothesise on the factors involved in a particular case (China post 1949, nineteenth-century Europe etc) and to explore the empirical validity of such hypotheses.

The examples given above were intended merely to illustrate some of the more commonly recognised problems in theory construction and to do so via instances drawn from population geography or demography. Figure 1.2 also lists five particular methodologies which will continue to have significance in geography. The remainder of this chapter is devoted to a brief review of the importance and potential of each of these methodologies for the study of population by geographers.

1 Empiricism

Both demography and population geography have been dominated by methodology that is implicitly empirical. Much attention has been given to the description of pattern, estimation, modelling and forecasting, but relatively less to interpretation or explanation. Statistical demographers have developed effective estimation procedures based on inadequate census or vital data whilst population geographers have made a significant contribution to the development of multi-regional demographic accounting systems (see Woods and Rees, 1986). Both require the construction of formal mathematical theories, but neither is immediately concerned with the explanation of demographic events. Geographers have also developed sophisticated techniques for the handling of spatially organised population data and especially their use in automated cartography.

Upon this basis of mapping and plotting, counting and describing rests a complex web of research that employ one or a number of other methodologies.

2 Positivism

When questions of a causal nature are addressed then some form of the positivist methodology is usually engaged. Positivism covers a number of variants on the same theme; but supposedly true propositions are first stated explicitly and then subjected to close scrutiny or testing. Although the shortcomings of this approach are now recognised (some were mentioned above) the aspirations of positivism applied to the human sciences are laudable yet beyond full realisation. In the study of population this methodology has been put to good effect especially in those circumstances that lend themselves to causal modelling of a multi-variate kind. For example, spatial variations in mortality, fertility and migration have been repeatedly treated in this way. Let us take variations in fertility as our illustration.

Fertility levels will be directly influenced by marital fertility, illegitimate fertility and the proportion of the female population married. Each of these components will be affected by other cultural, economic, social, political and demographic factors whose association with the dependent components of fertility may be hypothesised and identified empirically in specific circumstance via the co-variation of factors and components over spatial units. Some of the factors influencing marital fertility have already been suggested above (p 25). A substantial number of existing studies follow this line of argument with varying degrees of enlightenment (Teitelbaum, 1984; Anker, 1978; Bagozzi and Van Loo, 1978; see also Cadwallader, 1985 on migration).

However, the attendant problems are very obvious. First, hypothesis evaluation is usually accomplished through multiple-regression analysis which allows the signs and strengths of association to be specified, but depends on the factors being quantifiable. Second, associations may be unstable when the scale of analysis is altered. Third, factors are likely to be highly inter-dependent. Fourth, the co-variation of statistical attributes whilst providing a means of eliminating unlikely propositions does not give an opportunity to trace many of the more idiosyncratic processes involved in the creation of demographic behaviour and thus patterns. Pattern analysis by itself simply skims the surface of what is observable and measurable, it does not relate well to those aspects which may be intangible yet of deep significance. Despite these inherent drawbacks causal modelling has been used to some effect in population studies.

Several forms of positivism will remain important in population geography and demography, but now there is far more realism about the scope of the methodology, its reductionist nature and the potential of alternative approaches.

3, 4 & 5 Behaviouralism, Humanism and Structuralism

Figure 1.2 list three of these alternative approaches: behaviouralism, humanism and structuralism. Although they are all commonly used in human geography students of population have been less adventurous in experimenting with them than one might have expected. The behavioural approach has certainly proved of value for the study of migration decision making and probably has equal potential in unscrambling the sequence of decisions involved in marriage and family formation. Its chief failings relate to the preliminary assumption of rational human behaviour and the degree to which the attitudes and behaviour of individuals are thought to be determinable (see Cox and Golledge, 1981). This methodology does at least move the researcher closer to process recognition by focusing on motivation and not just pattern-form. Humanistic geography attempts a longer journey towards the experience of individuals via 'the understanding of human action through the study of meanings allocated to the element of the individual's life-world' (Johnston, 1983, 57). The researcher is obliged to imagine that which the individual experiences and to associate with it not his or her own meaning, but that of the object of study. The exercise is deliberately subjective, there is no attempt at prior hypothesis or explanation, understanding and interpretation are the goals. Amongst geographers in general the humanistic approach has been most widely developed in landscape studies, where its meaning is interpreted from the inhabitant's perspective, and in urban social geography, where participant observation offers some prospect of illumination (see Lowenthal 1961; Ley and Samuels, 1978; Jackson, 1983). These methods will also be appropriate to understand reactions to demographic events (arranged marriages, repeated pregnancies etc) through the feelings of those individuals who are directly involved.

How do Chinese peasants respond to the one-child family policy? What do unemployed Scots feel about migrating to London? What is the meaning of children for the Indian family? All of these questions may be tackled via the humanistic methodology. Yet once again there are certain difficulties. First, the results obtained by the researcher will not be replicable in the sense that positivist analysis may be reproduced. Second, the level of understanding obtained, whilst in one sense deep, will not transfer easily to the practical requirements of planners and forecasters. Thirdly, there is little agreement on exactly what procedures are to be followed, the investigator is left to his or her own intuition in the search for self-knowledge. These are important problems, yet more could be learnt from this approach which would shed light on the complicated precipitants of action.

The last methodology mentioned in Figure 1.2 is structuralism. Again there is no single agreed definition, rather the approach has been given a label that covers a number of different nuances of opinion. One central theme, however, refers to the significance of constraints on human action which both limit and enable certain behaviour patterns to emerge. In the Marxist tradition the structure of society is conditioned by the dominant mode of production and its related social formation (Carver, 1982). The behaviour of individuals is not a matter of free will, rather the form of society imposes constraints and obligations which severely limit choice. Behaviour is not determined, however. The organisation of the capitalist system, for example, requires workers to sell their labour power to the owners of the means of production for wages; it is based on the private ownership of property, and competition between production units, producers and employees. These relationships influence all other aspects of society, including demographic behaviour.

Marx was himself most concerned to refute the naturalistic representation of demographic systems which is to be found in Malthus's principle of population. In its place he advanced, but did not elaborate, the idea that each mode of production (ancient slavery, feudalism, capitalism, socialism, Asiatic) would have its own particular 'law of population'. Under capitalism there is need for an 'industrial reserve army' the relative size of which will be inversely related to wage rates. The 'surplus population' so generated will comprise the poorest section of society, the one subjected to the highest death rates, but the one having the highest fertility (Woods, 1983, 1986).

Although Marx's concept of historical materialism provides the most clearly articulated basis for a structuralist approach (as seen in Harvey, 1973) scholars of Marxian social theory are much preoccupied by the internal tensions that exist between the desires of individuals and the needs of society, and thus between the life that individuals make for themselves and that prescribed for them by wider social groupings. The concept of structuration has been developed to reconcile some of these tensions, in an intellectual sense at least. Giddens (1984) has argued for a combination of humanistic concepts with those of a more formalised structuralism so that individuals' actions would appear conditioned by the structural context within which they find themselves, but they would also be capable of acting to change that context and not merely reacting to it. Human agency would be given an important role in affecting 'life-world', but that set of relationships and meanings would also have a deeper significance emanating from the way in which society as a whole is ordered. The sceptic may see this merely as an accommodation between a naive idealism and an over-rigid materialism. Yet there are good reasons for a more positive response, especially from population geographers, who are concerned with both explanation and understanding, spatially articulated social constructs and individual decision making. The implications for empirical research remain unclear, as yet, but there are some slight signs of reorientation amongst some practitioners of the more single-minded approaches (see Woods, 1985, for examples).

BACKWARDS AND FORWARDS

The theories and methodologies sketched above provide examples of some of the ideas currently in vogue amongst social scientists whose concern is with population. Each of the approaches has advantages and limitations, none offer even the prospect of complete explanation or understanding yet none can be ignored. This plurality of approaches is only to be welcomed, it reflects a heterodoxy, a methodological garden of delights. There is no longer any need for eclecticism, choosing the approach to suit the question, a problem may be tackled using a number of approaches each one of which will shed new light from a different perspective. Nor is there need for competition between methodologies since each will work to its own epistemology. The three levels of theory outlined earlier will also tend to complement one another.

For the population geographer a number of very substantial problems do remain, however.

1. How should population geography be defined?
   This is an important, although tedious question. This chapter advocates the use of a narrow definition which emphasises the differences between human, urban and population geography. The last mentioned having an inner core which stresses spatial variations in mortality, fertility and migration, and thus the distribution and structure of population.
2. How may the influence of population geography be advanced within geography and in population studies in general?
   The latter will be accomplished when geographers adopt demographic techniques; persuade demographers, by example, of the value of post-positivist approaches; and reduce their preoccupation with migration. The former would be enhanced by the concentration of effort on particular questions (1 above), the development of consciously theoretical analyses; and the advance of modelling, forecasting and data handling/display techniques (the 'useful knowledge' approach).
3. What substantive questions remain unanswered in population studies that could be tackled by geographers?

> Obviously there are a number of issues but three will be mentioned here by way of illustration. First, there is as yet no comprehensive theory of migration which combines theoretical levels and methodologies (individuals - groups, causal models - humanistic interpretations etc). Yet here is an area in which empirical research by geographers has been fruitful and is most abundant. Second, the spatial variations in contemporary demographic regimes between cultural regions remain only partially understood whilst their historical development has largely been ignored outside Europe and America. Third, the application of forecasting methods via geographical data bases to issues of environmental management, housing policy and social planning has considerable unrealised potential to which population geographers can contribute their skills.

There need be no complacency regarding past achievements nor confusion of future purpose; population geographers have important contributions to make to their teaching subject, their research discipline and to closely related disciplines, especially demography.

REFERENCES

Anker, R. (1969) 'An Analysis of Fertility Differentials in Developing Countries', Review of Economics and Statistics, 60, 58-69

Bagozzi, R.P. and van Loo, H.F. (1978) 'Towards a General Theory of Fertility: A Causal Modelling Approach', Demography, 15, 301-319

Bähe, J. (1983), Bevölkerung in Globaler, Nationaler und Regionaler Sicht, Verlag Eugen Ulmer, Stuttgart

Becker, G.S. (1981) A Treatise on the Family, Harvard University Press, Cambridge, Mass

Berry, B.J.L. (1967) Geography of Market Centers and Retail Distribution, Prentice-Hall, Englewood Cliffs, NJ

Berry, B.J.L. and Horton, F.E. (eds.) (1970) Geographical Perspectives on Urban Systems, Prentice-Hall, Englewood Cliffs, NJ

Bulatao, R.A. and Lee, R.D. (eds.) (1983) Determinants of Fertility in Developing Countries (2 volumes), Academic Press, New York

Cadwallader, H. (1985), 'Structural Equation Models of Migration: An Example from the Upper Midwest USA', Environment and Planning A, 17, 101-113

Caldwell, J.C. (1982) Theory of Fertility Decline, Academic Press, London

Carver, T. (1982) Marx's Social Theory, Oxford University Press, Oxford
Clarke, J.I. (1965, second edition 1972) Population Geography, Pergamon, Oxford
Clarke, J.I. (1977) 'Population Geography', Progress in Human Geography, 1, 136-141
Clarke, J.I. (1978) 'Population Geography', Progress in Human Geography, 2, 163-169
Clarke, J.I. (1979) 'Population Geography', Progress in Human Geography, 3, 261-266
Clarke, J.I. (1980) 'Population Geography', Progress in Human Geography, 4, 385-391
Clarke, J.I. (1984) 'Geography, Demography and Population' in J.I. Clarke (ed.) Geography and Population, 1-10, Pergamon, Oxford
Cox, K.R. and Golledge, R.G. (eds.) (1981) Behavioral Problems in Geography Revisited, Methuen, London
Desbarats, J.H. (1981) 'Constrained Choice and Migration', Geografiska Annaler, 65B, 11-22
Easterlin, R.A. (1969) 'Towards a Socioeconomic Theory of Fertility: A Survey of Recent Research on Economic Factors in American Fertility', in S.J. Behrman et al. (eds.), Fertility and Family Planning: A World View, 127-157, Michigan University Press, Ann Arbor, Michigan
Easterlin, R.A. (1971) 'Does Human Fertility Adjust to the Environment?' American Economic Review, Papers and Proceedings, 61, 399-407
Easterlin, R.A. (1978) 'The Economics and Sociology of Fertility: A Synthesis' in C. Tilly (ed.), Historical Studies of Changing Fertility, 57-133, Princeton University Press, Princeton, NJ
Easterlin, R.A., Pollak, R.A. and Wachter, H.L. (1980) 'Toward a More General Economic Model of Fertility Determination: Endogenous Preferences and Natural Fertility' in R.A. Easterlin (ed.), Population and Economic Change in Developing Countries, 81-135, Chicago University Press, Chicago
Giddens, A. (1984) The Constitution of Society, Polity Press, Cambridge
Greenwood, H.J. (1981) Migration and Economic Growth in the United States, Academic Press, New York
Harvey, H.W. (1973) Social Justice and the City, Edward Arnold, London
Jackson, P. (1983) 'Principles and Problems of Participant Observation', Geografisk Annaler, 65B, 39-46
Johnston, R.J. (1983) Philosophy and Human Geography, Edward Arnold, London
Jones, H.R. (1981) A Population Geography, Harper and Row, London

Kosinski, L.A. (1984) 'The Roots of Population Geography' in J.I. Clarke (ed.), Population and Geography, 11-24, Pergamon, Oxford

Leibenstein, H. (1976) Beyond Economic Man, Harvard University Press, Cambridge, Mass

Ley, D. and Samuels, H.S. (eds.) (1978) Humanistic Geography, Croom Helm, London

Lieber, S.R. (1978) 'Place Utility and Migration', Geografiska Annaler, 60B, 16-27

Lowenthal, D. (1961) 'Geography, Experience and Imagination: Towards a Geographical Epistemology', Annals of the Association of American Geographers, 51, 241-260

Merton, R.K. (1967) On Theoretical Sociology, Free Press, New York

Morrill, R.L. and Pitts, F.R. (1967) 'Marriage, Migration and the Mean Information Field', Annals of the Association of American Geographers, 57, 401-422

Newman, J.L. and Matzke, G.E. (1984) Population: Patterns, Dynamics and Prospects, Prentice Hall, Englewood Cliffs, NJ

Noin, D. (1979) Géographie de la Population, Masson, Paris

Noin, D. (1984) 'Le Champ d'Etudes de la Démographie', Espace, Populations, Sociétés, 2 (2), 65-70

Notestein, F.W. (1945) 'Population: The Lone View' in T.H. Schultz (ed.) Food for the World, 36-57, Chicago University Press, Chicago

Perry, P.J. (1969) 'Working Class Isolation and Mobility in Rural Dorset, 1837-1936: A Study of Marriage Distances', Transactions, Institute of British Geographers, 46I, 115-135

Pryor, R.J. (1984) 'Methodological Problems in Population Geography' in J.I. Clarke (ed.), Geography and Population, 26-42, Pergamon, Oxford

Rhind, D. (ed.) (1983) A Census User's Handbook, Methuen, London

Skinner, Q. (ed.) (1985) The Return of Grand Theory in the Human Sciences, Cambridge University Press, Cambridge

Speare, A. (1971) 'A Cost Benefit Model of Rural to Urban Migration in Taiwan', Population Studies, 25I, 117-130

Teitelbaum, M.S. (1984) The British Fertility Decline, Princeton University Press, Princeton, NJ

Trewartha, G.T. (1953) 'A Case for Population Geography', Annals of the Association of American Geographers, 43, 71-97

Webber, H.J. (1972) Impact of Uncertainty on Location, MIT Press, Cambridge, Mass

White, P.E. (1981) 'Migration at the Micro-scale: Intraparochial Movements in Rural Normandy, 1946-54', Transactions, Institute of British Geographers, New Series, 6, 451-470

White, P.E. and Woods, R.I. (eds.) (1980) The Geographical Impact of Migration, Longman, London
Wolpert, J. (1964) 'The Decision Process in a Spatial Context', Annals of the Association of American Geographers, 54I, 537-558
Wolpert, J. (1965) 'Behavioural Aspects of the Decision to Migrate', Papers and Proceedings of the Regional Science Association, 15, 159-169
Wolpert, J. (1966) 'Migration as an Adjustment to Environmental Stress', Journal of Social Issues, 22, 92-102
Woods, R.I. (1979) Population Analysis in Geography, Longman, London
Woods, R.I. (1982) Theoretical Population Geography, Longman, London
Woods, R.I. (1983) 'On the Long-term Relationship Between Fertility and the Standard of Living', Genus, 39, 21-35
Woods, R.I. (1985) 'Population Studies', Progress in Human Geography, 9, 278-282
Woods, R.I. (1986) Malthus, Marx and Population Crises' in R.J. Johnston and P.J. Taylor (eds.), A World in Crisis, 127-149, Blackwell, Oxford
Woods, R.I. and Hinde, P.R.A. (1985) 'Nuptiality and Age at Marriage in Nineteenth-century England', Journal of Family History, 10, 119-144
Woods, R.I. and Rees, P.H. (eds.) (1986) Population Structures and Models, Allen and Unwin, London

Chapter Two

THE BRITISH AND UNITED STATES' CENSUSES OF POPULATION

J. C. Dewdney and D. Rhind

INTRODUCTION

Of all the sources of data used in modern population geography, none is richer than the national census, which provides a vast range of information on the population's size, distribution, structure, living conditions and movements. The practice of counting people has a lengthy history, which extends back over hundreds if not thousands of years, but the census as we know it today is essentially a product of the nineteenth and twentieth centuries. Earlier counts were carried out by authorities with some specific purpose in mind, such as taxation or military conscription, were usually confined to a particular section of the population - most commonly adult males - and were never published in full.

The modern census has been defined (United Nations, 1967) as "the total process of collecting, compiling, evaluating, analysing and publishing demographic, economic and social data pertaining, at a specified time, to all persons in a country or in a well-delimited part of a country." This implies an official, governmental activity, a comprehensive coverage of the entire population and the publication of a set of data referring to a specific point in time. Also implicit is the idea that a census should be one of a regular series, though this desirable situation has so far been achieved in only a minority of countries, including the United Kingdom. The modern census involves much more than a simple headcount and the publication of total numbers: it also involves the collection and publication of information on many of the characteristics or 'attributes' of the individual, such as his/her sex, age, marital status, birthplace, economic activity, social class, etc., together with data on the composition of the households among which the population is distributed. In many countries, including Britain, the census collects additional, non-demographic information appertaining to households, covering such matters as household tenure and amenities, number of rooms and type of dwelling. The published results are not confined to reporting the numbers of people or households with particular attributes, but are presented in the form of complex cross-tabulations of two or more characteristics, the cells of which are referred to as census 'variables'.

A few, small-scale national censuses were taken before the end of the eighteenth century (Iceland 1703, Sweden 1749, Norway 1769) but the first detailed census of a population of several millions was that of the United States in 1790. Britain and France followed suit in 1801 and first censuses were held in most European countries during the nineteenth century, the latest - and biggest - being that of the Russian Empire in 1897. Nineteenth century first censuses also occurred in 'Europe Overseas' (Canada, New Zealand 1851; Australia 1881), in several Latin American countries and in India (1881), but most of the 'Third World' was not covered until well into the twentieth century, often not until after World War II (Ghana 1948, Nigeria 1952, Zaire 1974).

THE BRITISH CENSUS

Origin and Development

The United Kingdom has one of the world's longest series of ten-yearly (decennial) censuses - eighteen in all - of which the first was held in 1801 and the most recent in 1981; only 1941, because of wartime conditions, is missing from the series. An additional 10 per cent sample census was held in 1966, but proposals for 'mid-term' samples in 1976 and 1986 were rejected on grounds of economy.

With the passage of time, the British census has become increasingly complex, collecting and publishing a progressively greater volume and range of information; major changes have also occurred in the organisation of the census operation and the geographical base for the collection and publication of the data.

The first four censuses (1801, 1811, 1821, 1831) recorded only the numbers of males and females in each house and family, with a simple occupational breakdown into five categories: persons in agriculture, trade, manufacturing, handicrafts and 'others'; details were also collected on the numbers of houses, families per house and whether the houses were vacant or occupied. A question on age was added in 1821.

The 1841 census recorded much the same information, but there were important organisational changes. The household became the basic unit of enumeration and the completion of the census form (or 'schedule') became the responsibility of the head of the household, assisted where necessary by the Enumerator. The local conduct of the census in England and Wales was transferred from the Parish Overseer to the

Registrar of Births, Marriages and Deaths, a post established when vital registration became compulsory in 1837. This change did not occur in Scotland until 1855 and the 1841 and 1851 censuses, like their predecessors, were organised by the official schoolmaster or 'other fit person' in each parish. By 1861, however, the whole of Great Britain had a uniform organisation comprising Enumerators responsible to local Registrars, who in turn reported to the Registrars-General of Scotland and of England and Wales. Under this system, the basic territorial unit became - and remains to this day - the Enumeration District (ED), a small piece of territory capable of being covered by one individual.

For the censuses of 1841 to 1901 (1861 to 1901 in Scotland) each Enumerator recorded the details from the census schedules in an 'enumerator's book', from which the data were extracted for tabulation. From 1911 onwards, punched cards were used for the recording and sorting of census data and the enumerators' books have contained only administrative details and a headcount of the population of the ED; the latter is used in the preparation of the population totals published in the Preliminary Reports (see below). The confidentiality of the books, as of the schedules themselves, is maintained for 100 years: those for the censuses of 1841 - 1881 inclusive have now been released and form an invaluable resource for detailed studies of the demographic and social structure of the population in that period.

The second half of the nineteenth century also saw a marked increase in the amount of information collected and published. The 1851 census was much more detailed than its predecessors: questions were asked not only on sex, age and occupation, but also on marital status, relationship to the head of household, birth place, nationality, education, economic activity (working, unemployed, retired) and whether the individual was deaf, dumb or blind. In 1891, additional questions were asked on the individual's employment status (employer, employee, self-employed), the ability to speak Welsh (in Wales) or Gaelic (in Scotland) and on the number of rooms in each household's accommodation. In 1911, the industry in which the individual worked was recorded, as well as his/her occupation, and there were questions on the fertility of married women.

From 1841 to 1901 inclusive, data collected at the ED level were aggregated for publication by Registration Sub-Districts, Districts and Counties, units which did not necessarily coincide with the contemporary administrative divisions. This source of confusion was removed in 1911, when the geographical base was altered to fit the pattern of Local Authority Areas. For the censuses of 1911 - 1971 inclusive, EDs were subdivisions of Wards and Civil Parishes and ED data could be aggregated thence to LAAs, which formed the basis of tabulations in the published volumes. The same applied in 1981 in England and Wales, but in Scotland EDs were based on postcode units. Changes in the boundaries of LAAs between successive censuses present problems in the study of intercensal change: in the case of population counts, at least, these can be overcome by reference to the tables published in the census volumes which record, in great detail, the boundary changes which have occurred since the previous census, together with the numbers of people affected. Special arrangements had to be made following the reorganisation of local government in 1974/75. These involved the publication of two sets of 1971 census volumes, one giving data for 1971 LAAs and a second containing 1971 data reaggregated to the new LAAs established in 1974/75.

The amount and nature of the data collected continued to change through successive censuses. In 1921, the fertility question was replaced by one on numbers of dependent children, and a new question was introduced on place of work, permitting the earliest studies of travel-to-work (commuting). The 1931 census omitted the questions on fertility/dependency, education and place of work but added those on unemployment and 'usual residence'.

The twenty-year intercensal period resulting from the absence of a 1941 census was a time of major demographic and social change and, for this reason, much additional information was sought in the 1951 census, making it the most detailed thus far. In addition to the questions previously asked on sex, age, marital status, relationship to head of household, birthplace, industry, occupation and employment status, those on fertility, education and place of work were reintroduced; questions on household amenities - the presence, absence or sharing of a piped water supply, bath or shower, kitchen sink, cooking stove and WC - were brought in for the first time. The 1961 census repeated virtually all the questions asked in 1951 and added new ones on household tenure (owner-occupied, rented from the local authority, privately rented, etc), on 'usual residents absent on census night' (permitting the calculation of de jure as well as de facto populations) and on 'address one year ago' (permitting detailed analysis of internal migration movements).

Following a period of heavy immigration, the 1971 census paid particular attention to that aspect. Although the question on nationality was dropped, each individual was required to state his/her parents' country of birth as well as his/her own and his/her date of entry into the UK if born abroad, questions subsequently attacked for their alleged racial undertones. Usual address five, as well as one, year ago was also demanded and the 1971 census was the first to ask for the individual's mode of travel to work (car, bus, train etc) and whether or not each household had the use of a car or cars. In 1981, the range of questions was considerably reduced, particularly as regards the contentious matter of immigration.

Both 'date of entry into the UK' and 'parents' birthplace' were dropped. The internal migration question was again confined to 'usual address one year ago', all fertility questions were dropped and the household amenities recorded were reduced to bath or shower and WC. On the other hand, several of the classifications used were expanded - those on marital status and mode of travel to work, for example. In terms of census output - the number of variables recorded and the detail of the cross-tabulations produced - the 1981 census has produced more information than any of its predecessors.

Collecting and Processing the Data

The legal basis. In Britain, as in most countries, the census is compulsory and is backed by the force of law. From 1801 to 1911, each census required a new Act of Parliament, but the 1921 and all subsequent censuses have been held under the provisions of the Census Act, 1920, the most important of which are as follows-

(i) The Act authorises the taking of a census of population under the direction of the appropriate Minister (in 1981 the Secretary of State for Social Services) at intervals of not less than five years. Thus no new Act of Parliament is required to conduct a census once five years have elapsed, and the operation is directed by an Order in Council.

(ii) The Order prescribes the questions to be asked which, under the terms of the Act, must be confined to the following topics: (a) name, sex, age; (b) occupation, profession, trade or employment; (c) nationality, race, birthplace, language; (d) place of abode, character of dwelling; (e) condition as to marriage, relation to head of family, issue born in marriage; (f) 'any other matter with respect to which it is desirable to obtain statistical information with a view to ascertaining the social or civil condition of the population'. This last provision might appear to give the census authorities the power to ask any question they wish, but in practice all proposals concerning the organisation of the census, the questions to be asked and the publication of the results are subject to the scrutiny of Parliament. The draft Census Order is debated in - and is not uncommonly amended by - both Houses of Parliament, and not until the final version is approved can the Minister issue the appropriate Census Regulations.

Particular attention is paid to the confidentiality of the information collected and no data may be released to any census user - which applies to other government departments as well as to local government officials and private organisations and individuals - in such a way that it would reveal the characteristics of an identifiable person or household. The complex 'confidentiality restraints' designed to meet this requirement have been described in detail elsewhere (Dewdney, 1981). The most important are that (i) all population data for areas with fewer than 25 inhabitants are 'suppressed' (i.e. not published), with the exception of total population numbers of males and females; (ii) all household data for areas containing fewer than eight households are suppressed, except for the number of households. These rules affect less than one per cent of ED's but much larger numbers of smaller areas such as postcode units or grid squares. Furthermore, the actual figures produced are 'adjusted' by the "the addition of a quasi-random pattern of +1, -1,0 to the individual cells".

Collecting the Data. Collecting, processing and publishing the data are the responsibilities of the census offices - in Scotland the General Register Office (GRO(S)), in Northern Ireland the Census Office of the Department of Finance, and in England and Wales the Office of Population Censuses and Surveys (OPCS). The last-named also has a general responsibility for ensuring collaboration among the three parts of the UK (which, nevertheless continue to collect and publish their data in slightly different ways) and with the Isle of Man and Channel Islands which, though technically outside the UK, hold censuses at the same time.

A large, for the most part temporary labour force is required to carry out the census. In 1981 this involved about 100 Census Supervisors, over 2000 Census Officers (with 6500 Assistant Census Officers) and about 130,000 Enumerators, together with a large temporary clerical staff in the census offices. Although the details of organisation were somewhat different in England and Wales, Scotland and Northern Ireland respectively, the post and duties of the Enumerator were common to all parts of the United Kingdom and it was the Enumerator who was responsible for the great bulk of the fieldwork, carried out over a period of four or five weeks.

As already indicated, the Enumeration District has long been the basic areal building block in the British Census. In 1981, Great Britain was divided into 130,047 EDs; thus, on average, an ED covered an area of about 1.8 sq km and contained a population of some 400, distributed among 150 households. In practice, the size of a ED and the number of people and households which it contained varied greatly according to local conditions of population distribution and density, with a predominance of large EDs with small populations in rural areas and small EDs with above-average populations in towns. EDs are formed, and their boundaries fixed, on the basis of two criteria: (i) they must fit within the boundaries of Wards, Civil Parishes and LAAs (in Scotland they are based on postcode units and sometimes require sub-division to produce Ward figures); (ii) they must represent roughly equal work-loads for the Enumerators, who are paid a lump sum for their efforts. Thus an ED may comprise a rural area of several sq km with a scattered population or, in towns, a collections of streets, a single street, part of a street or even a single block of flats.

The Enumerator is responsible for the following tasks:

(ii) the identification and listing, prior to the census, of all households in the ED;

(ii) the delivery to all identified households, two or three weeks before the census, of a publicity leaflet explaining its nature and purpose;

(iii) the delivery to each household, within a period of ten days before the census, of the census schedule;

(iv) collecting the schedules and checking them for errors and inconsistencies; with return visits to errant households where necessary;

(v) the provision of a preliminary count of the population of the ED for use in the preparation of the published Preliminary Report.

Clearly, the accuracy of the census depends on the care and conscientiousness with which the Enumerator carries out these tasks and the consistency with which he/she adheres to the instructions and definitions laid down by the census office. (OPCS 1981). Once collected and checked, the completed schedules are passed up through the chain of command to reach the census offices responsible for the production of the results.

Processing the data. Between the receipt of the census input - the completed schedules - and the emergence of the census output - the published results - there occurs a vast processing operation. In 1981, the resident population of Great Britain totalled 53.6 million, of whom 98.5 per cent lived in 19.5 million private households, each of which produced a household schedule (different schedules were used for the 0.8 million living in 'communal establishments'). The schedules for England and Wales and for Scotland (which differ in detail) are reproduced by Rhind (1983) and the questions used in 1971 and 1981 have been discussed in detail by Dewdney (1981, 1985) and others. This material is not repeated here: suffice it to say that the modern British census collects an enormous quantity of raw data.

In 1981, in addition to his/her name (which remains wholly confidential), eight pieces of information were required from each individual: sex, date of birth, marital status, relationship in household, whereabouts on census night, usual address, usual address one year ago and country of birth; all persons aged 16 or over were required, in addition, to state their 'degrees, professional and vocational qualifications' (if any) and 'whether working, retired, housewife, etc last week'; those at work had to provide the name and business of their employer, their occupation, their employment status and their means of transport to work; those no longer working had to provide the first three of these items in relation to their most recent job. In addition, six pieces of information were required from each household: number of rooms, tenure, sharing or lacking bath and/or WC, number of cars or vans available, and whether or not the household's accommodation was self-contained. The Enumerator had to provide a location reference, distinguish between households in permanent and temporary (caravans, etc) structures and to state whether or not the household shared a building with other(s) (Additional information was required in Scotland). The net result was a data set of approximately 1000 million pieces of information entered on some 19.5 million schedules.

The much smaller data sets from the first eleven censuses (1801-1901 inclusive) were aggregated and tabulated manually and from 1911 to 1961 the data were coded and punched onto cards for mechanical sorting and aggregation. In 1971 and 1981, however, data manipulation was carried out by computer and the new technology permitted the production of more numerous, more detailed and more complex tabulations than ever before. Data processing in these two censuses involved the coding of the items recorded on the census schedules for the production of computer tape files from which the many census tabulations were eventually produced. Not all the data were fully coded and stored: a considerable number of items were extracted from a 10% sample of the census schedules. Coding and storage involved two sets of operations: allocation to areal units depended on the locational references entered on each schedule by the Enumerator; allocation of household and population characteristics to the appropriate categories required coding of the answers to the various questions on the schedule.

The coding of 1000 million items was clearly a massive task. Much of the census schedule is now in the form of tick-boxes, to which code numbers are attached, and this part of the operation is relatively simple. Certain items, however, require transformation before coding: dates of birth (considered to produce more accurate answers than do requests for chronological age), for example. have to be translated into age (in years) at census date. Others, like 'business of employer' and 'occupation' have to be allocated to the categories of the Standard Industrial and Standard Occupational Classification respectively. Data involving the location of individuals require particularly laborious treatment; the answers to the migration question - 'usual address one year ago' - have to be allocated to the appropriate area and cross-referenced with 'usual address on census night' to produce the numbers of people who have moved between pairs of areas. In certain instances, items appearing in the census output are derived, during computer-processing of the primary coded data, from more than one item on the schedule. A prime example is the division of the population into 'socio-economic groups', which is arrived at by cross-referencing the answers on economic activity, industry, occupation and employment status. Such difficult items are coded only on a 10% sample basis, a method applied in 1981 to the questions on relationship in household, occupation, industry, workplace, means of transport to work and educational qualifications. Given the amount of work involved, the publication of the final results by September 1984 was a major achievement.

The Publication of the Census Results. As already indicated, the published results of the British census are by no means confined to simple counts of the numbers of persons or households in each of the categories used in coding and aggregating the raw data, but consist of sets of tables cross-referencing two or more populations or household attributes. Taking two examples at random there are figures for each areal unit showing the number of economically-active married females aged 20-24 working full-time in the week before the census, and the number of persons in private households in owner-occupied accommodation with an economically-active or retired head of household in socio-economic group 3 (self-employed professional workers). These 'census variables' normally appear as counts, from which 'derived variables' - such as proportions of all people or households - may be calculated by the census user.

The published results of the 1981 census appeared in a variety of forms, of which by far the most comprehensive were the traditional bound volumes and the Small Area Statistics (SAS).

The Published volumes. A set of bound volumes has been produced from every British census: in 1981 they numbered more than 1200 and contained some 18,000 pages of tables. Details may be obtained from the census offices' catalogues, from the annual HMSO catalogue, Government Publications and from the lists produced by Dewdney (1981, 1985) and others.

The volumes together present the full range of census data tabulated for Local Authority Areas and other "official" territorial units. In summary, they are of three main types:

Preliminary Reports, County Volumes and National Volumes.

(a) Preliminary Reports
The British census has a long tradition of producing, within a short period after census day and well in advance of the main results, a selection of preliminary data. In 1981, in a successful attempt to speed the production of the definitive volumes, this preliminary material was reduced to a minimum. Three Preliminary Reports were published, one each for England and Wales, Scotland and Northern Ireland, based on Enumerators' counts of the population present on census night. In addition, there was a special Preliminary Report on Towns: urban and rural areas giving population totals for urban areas as defined by their pre-1974 boundaries. Other early products were the two sets of Historical Tables, 1801-1981, comparing population counts in 1981 with those of all previous censuses.

(b) The County Reports (Regional Reports in Scotland)
These were the largest single block of 1981 census material - 148 volumes in all. These contain a standard set of tables, generally at the level of Local Authority Areas, covering the main demographic, economic, housing and household variables. In England and Wales, each report comprised two separately bound parts, Part 1 containing the 100% and Part 2 the 10% variables. In Scotland, there were four volumes for each region: Volumes 1 and 2 replicated most of the tables in the English and county reports, Volume 3 gave additional variables recorded only in Scotland and Volume 4 presented selected variables for lower levels of aggregation such as Wards, Civil Parishes and postcodes. The New Towns reports - two volumes each for Scotland and for England and Wales - contained the standard county report tables for those units. In Northern Ireland only one equivalent volume was produced, the Report for Belfast Local Government District, the rest of the data appearing in the 'national' volumes.

(c) National Volumes
This set included some volumes covering Great Britain - but never the United Kingdom - as a whole, and others for Scotland, Northern Ireland, England and Wales or Wales only, a situation reflecting the continuing compartmentalisation of the British census, which is carried out by a separate office in each "country". The national volumes fell into two main classes, those which replicated, usually at different levels of aggregation, the tables found in the county and regional reports, and those which presented additional material. In the first category were the two-part National Report, Great

Britain, the Scottish Summary, the Report for Wales and the Northern Ireland Summary Report.

A second set of national volumes, often referred to as the 'census topic' volumes, provided additional information on subjects not included or only partially covered in the county/regional reports and summary volumes. Each contained a valuable appendix indicating in which other volumes additional tabulations relating to its topic are to be found.

Several topics were dealt with in volumes covering the whole of Great Britain (Northern Ireland has its own set of topic volumes) - Sex, Age and Marital Status, Great Britain, Usual Residence, GB, Persons of Pensionable Age, GB, Communal Establishments, GB, and Qualified Manpower, GB. Others covered Great Britain but there was also a separate Scottish volume - Country of Birth, Economic Activity, Workplace and Transport to Work. In the cases of Housing and Households and Household and Family Composition, there were separate volumes for Scotland and for England and Wales. Migration data occupied no fewer than 24 volumes: National Migration, Great Britain Part 1 (100%) and Part 2 (10%), equivalent Regional Migration reports for each of the nine standard regions of England and Wales and four Scottish Migration volumes, 1 and 2 containing the 100% and 3 and 4 of the 10% data. Finally there were two special volumes on language - Welsh Language in Wales and the Scottish Gaelic Report.

(d) 'Key Statistics' volumes

The sheer scale of the published output presents problems for the user requiring quick and easy access to a basic piece of information for a particular area. To supply this need, OPCS and GRO(S) jointly produced the summary volume Key Statistics for Local Authorities, Great Britain, which presented, in 19 tables, some 120 variables for the 459 local government districts, most of them in percentage form. A further six 'key statistics' volumes were produced for specially identified urban areas - Key Statistics for Urban Areas, Great Britain, Key Statistics for Localities (Scotland) and four regional volumes, Key Statistics for Urban Areas, North, Midlands, South East and South West and Wales.

(e) Other Publications

An innovation of the 1981 census was the printing of selected variables many of them in percentage form, in booklet series - County Monitors, Ward and Civil Parish Monitors and Parliamentary Constituency Monitors. Additional information on economic activity is available from a set of Economic Activity Booklets, published in microfiche.

In addition to the data themselves, the census offices have produced a much greater range of explanatory documentation than on any previous occasion. Too extensive to be listed here, this material includes Census Monitor series and some 200 User Guides, details of which can be obtained from OPCS.

The Small Area Statistics (SAS). SAS have now replaced the tables in the published volumes as the main resource for research on the British population. Their development began in 1961 when, for the first time, data were made available for EDs in selected areas, and continued in 1966, when data were provided for all EDs. The supply of SAS was systematised in 1971 by their provision in a standard format of 1571 variables arranged in 28 tables. In 1981 they consisted of some 4300 variables (5000 in Scotland) arranged in 52 tables, organised into 10 pages. The layouts for 1971 and 1981 may be obtained from OPCS and have been reproduced by Rhind (1983) and Dewdney (1981, 1985).

SAS are available as printed copies (the 'pages' referred to above), as microfiche, on microfilm or on magnetic tape for computer manipulation, which is now the form most commonly used. Each variable has its unique cell number. Data can be extracted and manipulated by reference to these cell numbers, which can be used to construct 'derived variables'. As an example, cell 50 is 'total residents' and cell 112 is 'married female residents aged 30-34'. Thus the latter can be expressed as a percentage of the former by the specification x 100. SAS are available for Enumeration Districts, Wards, Civil Parishes, Postcode sectors (Scotland only), Local Authority Areas and Parliamentary Constituences. Space does not permit a description of the contents of the SAS tables: suffice it to say that they would appear to contain virtually every meaningful classification and crosstabulation which it is possible to produce from the raw data recorded on the census schedules.

With the appearance of SAS in a form suitable for manipulation by computer, much attention has been devoted to the preparation of statistical packages for use with these data. The most significant of these is SASPAC, produced specifically for the 1981 SAS by the Universities of Durham and Edinburgh under contract to the Local Authorities Management Services and Computer Committee (LAMSAC). SASPAC can be run on a variety of different computers, produces straightforward analyses of the data and prepares data for input to more complex statistical and mapping packages. Among the latter, one of the most widely used is GIMMS, produced by Gimms Ltd. of Edinburgh. Further development along these lines seems likely before the next British census, scheduled for April 1991.

THE AMERICAN CENSUS

Origin and Development

The most fundamental aspect of the American Census is that its origins are bound up with the formation of the nation state. It is enshrined in the Constitution. The delegates to the Constitutional Convention of 1787 decided that both state apportionments for the House of Representatives and direct taxation should be based upon population distribution. By combining the two, the legislators sought to achieve a balance between pressures for over- and under- enumeration. Article I, Section 2 of the Constitution decrees that a Census should be held at least once every 10 years commencing within three years of the first meeting of Congress. Even now, the primary driving-force behind the Census remains apportionment, evidenced by the Court actions taken by north eastern cities against the Bureau of Census after 1980 on the basis of alleged under-enumeration; other uses of Census data have, of course, burgeoned since 1790.

The first national Census actually post-dated some 38 earlier Censuses carried out in the Colonies, the earliest of which was held in Virginia in 1624, and most of which were carried out at the request of the British Board of Trade. Though Madison and others in the first Congress argued for a wider information-collecting role, the 1790 Census recorded only the name of the head of family, and numbers therein. These were sub-divided into free white males (above and below 16 years of age), free white females, all other free persons, and slaves. No count was made of Indians in the hinterlands, since they did not contribute to revenue and, for apportionment purposes, five slaves were to be counted as the equivalent of three persons! The actual job of carrying out this Census was mainly in the hands of the Federal Marshals in each of the then 16 States, who hired enumerators. Beginning on August 2 1790, the Census was completed in nine months, the last results being turned in by March 1792. To carry out the Census, much local enterprise was required - many enumerators had to provide their own paper. It is not surprising therefore that the 3.9 million individuals recorded was generally accepted to be a significant undercount of the true population.

The Censuses held immediately thereafter introduced a number of important innovations. The 1820 Census, for instance, contained questions on whether employment was in agriculture, commerce or manufacturing (though the results were found to be very inaccurate). Printed schedules appeared in 1830 and a centralised office for running the Census was created for that purpose in 1840 - though it was to be disbanded once the reports were issued. By that time, the breadth of information collected had increased dramatically, the Census Act of 1840 instructing the office to collect "all such information in relation to mines, agriculture, commerce, manufactures and schools as will exhibit a full view of the pursuits, industry, education and resources of the country". From 1850 to 1870, the Marshals were instructed not only to use the individual rather than the household as the unit of enumeration but also to produce reports for civil divisions which they would create and were to be known as counties.

By 1870, the rapid growth in information collected had reached astonishing proportions. This had several effects, notably that the Marshals were swamped with information, were unable to validate it properly and produced the reports only when the data were obsolete. The Census Act of 1880 did, in fact, institute a temporary Census office within the Department of the Interior and made provision for specially appointed enumerators and supervisors in place of the Marshals. The Act was also the first to prohibit enumerators from disclosing any information collected in the Census. An innovation of great moment in the 1890 Census was the introduction of Hollerith's punched card tabulating machines, but perhaps just as significant was the establishment of a permanent Census office in 1902. Three years later this moved to the Department of Commerce and a Census geographer was appointed. As well as being the first taken by this new organisation, the 1910 Census also included a small experiment which, whilst soon abandoned, had immense repercussions for contemporary American Censuses - the use of the public mail service to deliver questionnaires.

The 1940 Census, however, provided the prototype for today's Censuses: it included the first Census of housing and the first substantial recording of employment, unemployment and migration. Perhaps most important, however, it introduced the concept of sampling to reduce the amount of work in subsequent data processing. By 1960, only a few of the population items were sought from all of the respondents and much of the distribution and collection of Census forms was done by mail. That year was also the first in which all Census tabulations were made by computer, the Census Bureau having installed a UNIVAC 1 - arguably the first commercial computer - in the late 1940s. Other important technical developments pioneered or advanced by the Bureau include FOSDIC or Film Optical Sensing Device for Input to Computers. Special questionnaire forms were designed, filled in by the respondent, microfilmed and then scanned by computer to obviate key-punching.

As in Britain (one year later), the 1970 Census saw a substantial increase in the number of statistics tabulated and particularly in those for small geographic areas: details were released for 1.7 million city blocks and 35,000 Census tracts. Only a few basic population and housing questions were asked on a 100% sample basis, most being collected on a 15% or 5% basis. That Census built heavily upon the 1960 experience of use of the public mail. Separate FOSDIC-readable questionnaires were used for rural and for urban areas and 60% of the total population in large metropolitan areas received questionnaires by mail and returned them after completion by the same mechanism. Enumerators contacted only non-responding households or those returning incomplete forms. Elsewhere in the country, forms were sent by mail and collected by enumerators.

THE 1980 CENSUS

The Census Context

The 1980 Census cost nearly $1 billion, compared with $222 million for that in 1970. This enormous enterprise was planned from 1973 onwards in the context of several important factors, including:

- a widening range of user needs, especially from commercial agencies for small area data
- changes in life style which complicate taking the Census, e.g. the substantial number of illegal immigrants, new forms of household have been created, and living in remote spots had, if anything, become more common
- the labour market changes (working housewives, etc.) had resulted in fewer skilled and conscientious individuals being available as enumerators. It had also ensured that far more calls would need to be made in the evenings to collect forms
- the racial and ethnic variations in the US ensured that enumerators in homogeneous areas had to be from the same background as those enumerated, thereby complicating still further the choice of enumerators.

The Mail Out-Mail Back Census

Some 90% of the US population (but only those in 50% of the land area) were enumerated in 1980 using a mail out-mail back Census procedure. The claimed advantages of this procedure are substantial: fewer enumerators are required, hence saving costs, the household has time to reflect on answers and provide them more accurately and it offers greater privacy to the respondent. In reality, however, other factors also led to its adoption. These included the difficulties of obtaining suitable enumerators, the remoteness of some settlements, and the relatively poor quality of the US map base, at least when compared to that in Britain. The last factor is often unappreciated by UK residents: the largest scale of complete, consistent map coverage in mainland Britain is 1/10,000, with

70% of the area covered by 1/1250 and 1/2500 scale maps which are continuously revised. Hence, purchase of the SUSI (Supply of Unpublished Survey Information) in Britain should produce maps which are no more than a few months out-of-date. By contrast, the largest scale continuous coverage in the USA is at 1/62,500 scale and many of these are up to 40 years out-of-date. In these circumstances, mailing lists often constitute the most up-to-date detailed population geography. Despite this, the best available maps of the USA were compiled for each one of the 300,000 Enumeration Districts for use in the field (below).

Mailing lists for 1980 were generally obtained from commercial sources, though several for the same area were sometimes merged and the results checked by the postal service in mid-1979. Prior to the Census, an enumerator systematically checked all the addresses in an area, adding missing ones to the mailing list. The numbers of households in different areas from the city block upwards were thus validated by comparison with local water, gas and electricity and administrative records: over 32,000 of the 39,500 local units of government participated in this checking process of the address lists. Other checks and up-dates continued almost up to Census day on April 1st.

It is worth stressing that, to make the mailing operation a success, enormous publicity efforts were made: Census messages were placed in employees' pay packets, on grocery bags, in the media and were even given out in schools. A response rate of 74% was achieved in the mail out-mail back operation (GAO 1982).

The Geography of the US Census

The geography of Census-taking. Attaching a geographical reference to a Census record in the 1980 Census required three products produced by the Geography Division (Marx 1984). These were:

(i) Maps, showing named streets and boundaries of enumeration districts, Census tracts, townships, counties, cities and major physical features (such as railways) to aid position fixing. The maps were used in urban areas primarily by enumerators when checking on non-respondents.

(ii) Geographic Base Files (GBF)/Dual Independent Map Encoding (DIME) files. These encode the topological relationships between streets and between the postal address and the postal address and the block (see Figure 2.1). Given a GBF/DIME file, automated matching of Census records containing an individual postal address with a section of a street and hence

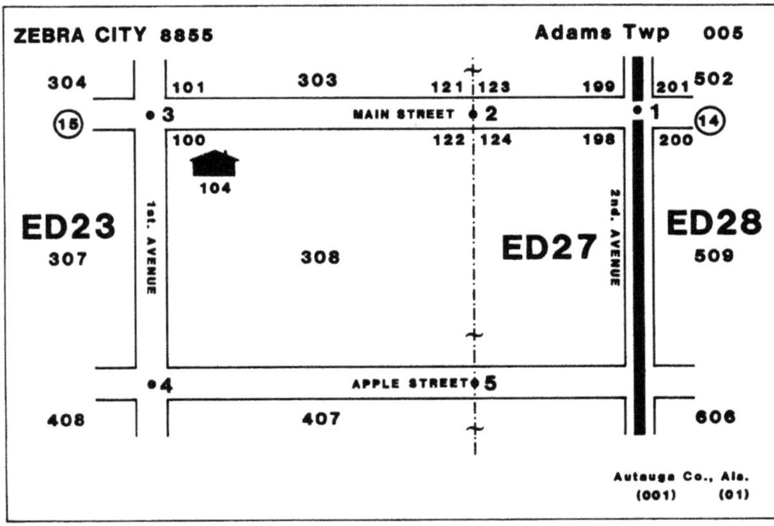

Street Name	From Node	To Node	Place	Tract	Left Block	Right Block
MAIN ST.	3	2	8855	15	303 101-121	308 100-122
MAIN ST.	2	1	-	15	303 123-199	308 124-198

Fig. 2.1 Portion of a US Census map showing nodes, streets, blocks, enumeration districts and postal numbers (top); and section of part of the equivalent GBF/DIME file (from Marx 1984).

with blocks is generally possible. Moreover, the structure of the DIME file permits automatic checking that much of the street network is complete - or at least that what is recorded is consistent. This is vital since GBF/DIME files were created in collaborative projects between the Census Bureau and 300 local agencies.

It is worth noting that this only provides implicit geographical referencing - the standard GBF/DIME file contains relationships rather than geometry of streets, etc. However, for the 1% of the land area which includes 60% of the population, the latitude and longitude of all the street nodes have been converted into computer form and plans are in hand to extend this for the 1990 Census (see below).

(iii) the Master Reference File. For each State, this contains the relationship between the geographic units (street sections, etc.) in the GBF/DIME file with the areal units for which the Census results are to be tabulated (hence it is the US equivalent of the OPCS Area Master File which links EDs to a variety of Health Service areas, etc.). For 1980, this was created by a massive clerical effort to encode the 2.5 million block numbers on the Census maps together with the enumeration district numbers, Census tract numbers, place names and codes, etc.

Given these three elements, the capability for automatic tabulation of results for an almost infinite variety of geographic bases exists.

The Geography of Census Reporting. The tabulation units for which 1980 Census results were produced were:

Table 2.1: Tabulation Units Recognised in the 1980 Census (from Marx 1984)

Political Areas	Statistical Areas
United States	Regions (4)
States & State Equivalents (57)	Divisions (9)
States (50)	Standard Consolidated
D.C. (1)	Statistical Areas - SCSA (17)
Outlying Areas (6)	Standard Metropolitan
Countries, Parishes & Other	Statistical Areas - SMSA (323)
County Equivalents (3,231)	Urbanised Areas - UA (373)
Minor Civil Divisions - MCD (30,491)	Census County Divisions - CCD (5,512)
Incorporated Places (19,176)	Unorganised Territories (274)
American Indian Reservations (275)	Census Designated Places - CDP (3,733)
Indian Subreservation Areas (228)	Census Tracts (43,383)
Alaska Native Villages (209)	Block Numbering Areas - BNA (3,404)
Congressional Districts - CD (435)	Enumeration Districts - ED (102,235)
Election Precincts (36,361)	Block Groups - BG (156,163)
[In 23 participating States]	(Tabulated parts - 200,043)
School Districts (15,850)	Blocks (2,473,679)
Neighbourhoods (≅ 27,750)	(Tabulated parts - 2,545,416)
ZIP Codes (≅ 37,000)	

Two of these tabulation units are worth examining in more detail. The block is often a very small tabulation unit indeed, and only selected statistics from the 100% Census sample are published for it. Block data are, however, published for all urbanised areas and all incorporated places of 10,000 or more people. In addition, other areas can contract for the Bureau to produce block data - the totality of the States of New York, Rhode Island, Virginia, Georgia and Mississippi are available in this form. The Census tract is an unchanging area, hence facilitating comparison through time. Unlike the areas of the same name introduced in 1981 British Census results, this is (in general) initially defined on the ground by a Local Census Statistical Areas Committee as a homogeneous area of population and housing; other government and administration is urged to use these areas. The concept and naming of the 'tract' is primarily due to Laidlaw whose assembly of historical and current data for small areas was threatened when New York changed the boundaries of their assembly districts (and hence of Census reporting units) in 1905. Census tracts exist within all the Standard Metropolitan Statistical Areas and some other areas: their average population is 4,000, with a range of 2,500 to 8,000 people.

The 1980 Census Questions

The 1980 Census forms are given in Kaplan and Van Valey (1980). Questions covering 7 population subjects and 12 housing subjects were asked of every household. The population items were name, relationship to household, age, 'race', sex, marital status and Spanish origin. The remaining questions dealt with 26 population and 20 other housing items but were collected on a sample basis as previously described: in total, 79% of households got the short (100% only) forms. Apart from the complexity of the race and ethnic classification (Kaplan and van Valey 1980, p.282-3), perhaps the most interesting items are those collected on the sample basis. Those in the population section include education, place of birth, citizenship and year of immigration, current language, ancestry, residence and activity 5 years previously, veteran status, disability, number of children ever born, marital history, employment status, place of work and journey to work, the industry, occupation and class of worker, number of weeks worked in 1979 and income. The last of these has appeared in each Census since 1940 and was specifically included because it is needed for Federal fund allocation formulae. The 100% sample housing items included access to the housing unit, plumbing (household amenities in the British Census), number of rooms, tenure, property value for owner occupied units and, otherwise, monthly rent, and vacancy status. The sample-based questions included building type, number of stories and presence of an elevator, source of water and sewage disposal, year in which the building was constructed, the year in which the present occupant moved in, heating equipment, fuels used, the cost of utilities and fuel, kitchen facilities, number of bedrooms and of bathrooms, presence of telephone and of air conditioning and the ownership and use of cars, vans and light trucks. Clearly this range of questions is somewhat wider than in the British Census.

Census Results

The single report from the first US Census amounted to only 56 pages: the equivalent number of printed pages from the four Censuses from 1950 to 1980 were 50,000, 100,000, 200,000 and (about) 300,000. In addition, from 1960 onwards, increasing volumes of data have been made available in computer-readable form on magnetic tape.

The results available differ to some extent, depending on the transmission medium adopted. Hence the paper reports began with Preliminary Population and Housing Unit Counts, superseded by equivalent Final Counts and, in addition, basic and some extra counts are provided in the PHC series, individual reports being produced for block and Census tract level within each SMSA and another for governmental units in each State. In addition, a whole series of population and household reports describe the basic data for a variety of areal units; whilst cross-tabulations do occur in these reports, they are much less common than in their British counterparts. The same is true (although to a lesser extent) for the subject reports, mostly produced at national level.

Computer tapes contain much more detail both in terms of cross-tabulations and geographical detail - hence one tape (STF1C or Standard Tape File 1 for Counties contains well over one hundred cells or variables for each of the 3,000+ counties in the USA. Census of Population and Housing (1980) provides details of the STF files which are readily obtainable. It is worth pointing out that, unlike the British policy of charging by data per areal unit, the US Bureau of Census charges only the cost of copying the data. Hence a tape full of data from OPCS (say EDs in a county) could cost £2,000 or more, whilst the maximum charge for the 'public domain' US data appears to be about $100.

One other product produced in the US is worth examining: the Public Use Microdata Sample tape files which contain a small sample of unidentified individual and household records, each containing virtually all the details other than name and address for those people or households selected: only States or groups of counties are identified. Norris (1983) pointed out the advantages of such data in permitting the user both to create his own cross-tabulations and also to avoid the ecological fallacy: he urged that these data should be made available for Britain and discussions are now in hand between OPCS and the Economic and Social Research Council to see if an equivalent of these data can be produced.

Confidentiality

Since 1910, it has been the practice that a Presidential proclamation is given of confidentiality of Census information. The Carter proclamation of 1979, for instance, included:

> "....By law, individual information collected will not be used for purposes of taxation, investigation, or regulation, or in connection with military or jury service, the compulsion of school attendance, the regulation of immigration or with the enforcement of any national, State or local law or Ordinance".

Whilst this is somewhat more detailed than are the terms of the 1920 Census Act in Britain, the individual questionnaire responses are retained as confidential for a shorter period (72 years compared with 100 in Britain).

In addition to the total prohibition of giving of individual's records to other agencies, the Census Bureau is charged with ensuring that no individual's characteristics can be obtained from the published results. The Bureau achieves this in a different manner to the British - fewer cross-tabulations are produced and those which are are often only produced for areas with more than a given population size (e.g. the Summary Tape File 5 is only available for States, SMSAs and SCSAs). Hence only simple counts are given at the block level (averaging about 70 people).

Where detailed data are presented, data suppression techniques are applied, but they differ somewhat from the British approach. Suppression in the STF1 tapes can become quite complex: in general, figures are set to zero if fewer than 15 people or 5 households exist in a table other than those of basic counts. But these thresholds can also be applied to numbers of Spanish origin or individual racial group in selected tables and complementary suppression (suppression on two or more bases, e.g. 'five owners and five renters') is used in some data.

Accuracy

US Censuses have characteristically undercounted the population though in general the discrepancy has been diminishing in size from 3.3% in 1950 to 2.5% in 1970. However, this undercount varies greatly between social and ethnic groups and in different areas of the country - in particular, the error is greater in regard to Hispanics than for most other groups. Clearly all this has major significance for apportionment at State level but the need for adjustment to obviate the undercount is present even down to the 39,000 sub-State units which gain from revenue-sharing programmes based on population numbers. Refining the accuracy can of course be achieved by various supplementary, post-census checks, though these are often expensive. The Bureau spent $11 million in 1970 to reduce the undercount by 0.2%. In 1980, an estimated $342 million was spent in total (GAO 1982) on reducing the undercount.

THE 1990 CENSUS

US planning for a Census now begins about 7 years prior to a Census and much reassessment of the 1980 Census was undertaken in the period 1981-83. The General Accounting Office of the Comptroller General of the US, for instance, carried out two investigations (GAO 1982, 1983) on how the 1990 Census could most efficiently and economically be carried out by modifying 1980 procedures: the Department of Commerce responses to the many detailed recommendations were also printed in these reports. Bounpane (1983) has written a 'look ahead' from within the Bureau which describes the possible modifications being examined. One of the most significant of these hinges on the introduction of the TIGER (Topologically Integrated Geographically Encoding and Referencing) System, a computer system to combine maps, addresses and Census geographical areas into one data base. This has two objectives - to reduce the still high levels of manual effort at the Census preparation (c.f. the highly automated Census tabulation) stage and to reduce the errors and inconsistencies that arose in 1980 from having separate maps, GBF/DIME files and Master Reference Files. Broome (1984) has given a description of the design of this system. To make this function satisfactorily, a collaborative programme with the US Geological Survey has been set up to create and convert into computer form a totally new set of 1/100,000 scale maps of the USA by 1987.

THE BRITISH AND US CENSUS COMPARED

Whilst they have many superficial similarities - they are large and complex operations, orientated towards an accurate count of the population, held every ten years and whose results are fundamental for much planning and resource allocation - there are many differences between the approaches used for Census taking in the two countries.

Perhaps the most important of these differences is the method of distributing and collecting the questionnaires, the way in which sampling is carried out, the type and range of questions asked, the data encoding, validation and tabulation procedures used, the nature of the basic Census geography recorded and the cost of obtaining the results. Naturally, many of these arise from differences in the size of the countries, the terrain to be covered, and the legislative and social context in which the Censuses are carried out. It may be that the internationally exploited and rapid developments in new information technology and the pressures upon Census agencies to reduce costs (e.g. GAO 1982) will lead to more commonality but each Census is essentially context-specific.

REFERENCES

Bounpane, P.A. (1983) 'The Census Bureau Looks to 1990' American Demographics, October, 1983, 1-8

Broome, F.R. (1984) 'Tiger Preliminary Design and Structure Overview,' Paper Presented to Assoc.Amer.Geographers Annual Meeting in Washington D.C

Census of Population and Housing (1980) Summary Tape File 1 Technical Documentation, Data User Services Division, Bureau of Census, Washington D.C., 2nd Edition 1982 (equivalent publications cover SFT 2 to 5)

Denham, C. (1980) 'The geography of the census, 1971 and 1981', Population Trends, 19, 6-12

GAO, (1982) A $4 billion Census in 1990? Timely decisions on alternatives to 1980 procedures can save millions, General Accounting Office, Washington D.C., GGD-82-13

GAO, (1983) The Census Bureau needs to plan now for a more automated 1990 decennial census, General Accounting Office, Washington D.C., GGD-83-10

Kaplan, C.P. and Van Valey, T.L., (1980) (eds.), Census 80: Continuing the Factfinder Tradition, Bureau of Census, Washington D.C., 490 pp

Marx, R.W. (1984) Developing an Integrated Cartographic/Geographic Data Base for the US Bureau of Census, International Bureau of Census Memorandum

Morgan, C., and Denham, 'Census Small Area Statistics (SAS): Measuring Change and Spatial Variation', Population Trends 28, 12-17

Norris, P. (1983) 'Microdata From the British Census', in D. W. Rhind (eds.), A Census Users Handbook, Methuen, London, pp.301-19

OPCS, (1981) Census 1981: Definitions, Great Britain, (CEN 81 DEF), HMSO, London

Rhind, D.W. (1983) (ed.), A Census Users' Handbook, London, Methuen

UN, (1967) Principles and Recommendations for the 1970 Population Censuses, UN Statistical Papers, Series M, No.44

Chapter Three

FERTILITY PATTERNS IN THE MODERN WORLD

J. Coward

Like most population characteristics, the analysis of fertility patterns can be investigated at many different scales of study and this chapter therefore examines fertility patterns with examples from a variety of points along the scale continuum. At the more macro level, fertility will be assessed for the countries and sub continental areas of the world, emphasising the considerable degree of contemporary variation and the diversity of factors that influence these differentials. At a somewhat finer level of analysis, fertility variations in the British Isles will also be examined. Here, several levels of aggregation are considered, focussing on the constituent countries of the British Isles, regional and sub regional variations within each of these areas and also fertility differences at the district level in England and Wales. Prior to these specific examples, however, the chapter discusses some general issues concerning the geographical study of fertility patterns, examining the possibilities and limitations of such studies, data availability and the nature of theory.

GENERAL ISSUES

Although the field of population geography has expanded considerably in recent decades, the analysis of fertility patterns by geographers remains a relatively neglected area of study. In many respects this is a surprising feature because there appears to be considerable scope and possibility for the geographical study of fertility. Firstly, for example, spatial variations in fertility are generally discernible for all the gradations of the scale continuum. Secondly, many countries now have at least the basic data sources necessary to investigate spatial variations in fertility. Thirdly, the inter-disciplinary nature of fertility studies, in which a wide range of biological, demographic, socio-economic, socio-psychological, political and environmental factors can be of potential explanatory importance, offers much opportunity to demonstrate that the geographical approach is one based on the science (or art) of synthesis. Fourthly, while real variations in fertility can be an important study in themselves, they can also be utilised as indicators of wider issues, such as variations in social change, or related to more applied population issues such as variations in overall population change, distribution and structure and also the demand for education, health and service provision

This neglect of fertility studies by geographers was emphasised by Trewartha (1953) over thirty years ago, but more recent surveys of the literature, such as Clarke (1977), Jones (1981, 1984), Schnell and Monmonier (1983), Rees (1981) and Woods (1979, 1982) come to much the same conclusion. Jones (1984), for example, states that the neglect is "serious and indefensible", but has also suggested that it can be explained by fertility studies being perceived as less "geographical" (presumably reflecting attributes of the physical environment or such spatial attributes as the direct effects of distance) than, for example, studies of mortality or migration (Jones 1981). It is certainly the case that in most circumstances the physical environment plays little or no part in directly influencing fertility patterns, although, at broad scales of study, the environment can effect patterns of health and diet which can, in turn, influence fertility characteristics. On the other hand, the role of distance can sometimes play a crucial role, as in, for example, distance decay relationships or in the geographical diffusion of certain ideas and information.

Other reasons, however, may also explain the neglect of fertility studies. These could include, for example, the lack of substantial variation in fertility in many developed countries at the broad regional scale; the generally greater relative importance of migration in influencing population change at regional and sub-regional scales and particular problems relating to data availability and reliability. In terms of data problems, the study of fertility shares many of the possibilities and limitations of data availability common to any geographical study of population. Thus some of the limitations include unreliability in vital registration and census sources due to under-reporting, under-enumeration or mis-statement of information (particularly amongst statistically under-developed countries); unavailability of data regarding certain key explanatory variables and, until recently, little data available at the more detailed scales of analysis. However, considerable progress has been made recently regarding data collection and publication and this has greatly added to the scope of geographical studies of fertility. Thus, at an international level, various agencies are concerned with estimating and publishing population numbers and vital rates (particularly the United Nations and the Population Reference Bureau) whilst, amongst developed countries in particular, greater attention has been paid to providing more detailed vital statistics, detailed census information for small areas and record linkage via geographical information systems. Furthermore, the increasing use of sample surveys concentrating on fertility or including fertility questions can again prove useful in the study of spatial patterns of fertility - for example the recent work of the World Fertility Survey and, in many countries, greater reliance on household surveys. The advantage of such surveys is that the researcher often has access to information at the individual level and that the range of information is much wider than that associated with the standard census. On the other hand these surveys are designed to be representative of the whole country and reliable information at the sub-national scale is probably only available at a broad regional level. The General Household Survey in Britain, for example, provides much information on patterns of family formation and family size at the national level, although problems of sample size prohibit spatial studies at scales finer than that of the standard region. Thus while it can be seen that population studies will always be hampered by "data problems", the range and quality of the data sources available for fertility studies hardly appears of relevance in attempting to understand geographers neglect of this subject.

Finally, the role of theory in spatial studies of fertility can be briefly assessed. There are two main sources of theory that are applicable to geographical studies of fertility involving, on the one hand, a framework that is either explicitly spatial, such as diffusion theory or distance decay relationships, or, on the other hand, applying theories developed by other disciplines (often used in explaining changes over time) in a spatial setting. These two types are often combined in a single approach, such as the investigation of the effects of the spatial diffusion of modernisation upon resulting spatial patterns. Of the first type, diffusion theory provides a tempting area of study but one which proves difficult to investigate. Thus it might be expected that ideas and information regarding fertility would spread (at a variety of spatial scales) from certain core areas to outlying areas and thus influence in varying degrees the resulting patterns of fertility. Jagielski (1980), for example, examines how the spread of low fertility from urban areas to surrounding rural areas can reflect the diffusion of the small family norm. However, as Jones (1981) suggests, it is often difficult to envisage what actually constitutes the process of diffusion. It is unlikely that knowledge of birth control practices would diffuse in this way because a variety of historical and anthropological studies indicate that such knowledge is normally indigenously generated. On the other hand, the spread of information concerning the availability of certain artificial forms of birth control might have more application to fertility studies in certain contexts, but even then the key factor is generally the varying motivations of individuals concerning the control of fertility rather than knowledge about family limitation in itself. However,rather than the spread of information regarding fertility limitation, the wider process of the diffusion of culture may well have important repercussions on changing spatial patterns of fertility. Thus Caldwell's (1977) example of the influence and diffusion of westernisation in Nigeria would seem to be particularly relevant when examining the changing spatial patterns in that country. Diffusion of information concerning fertility behaviour remains a tempting yet tantalising approach to the study of spatial fertility. On the other hand, the importance of distance decay relationships seems more clear cut, particularly in a local context where there are variations in the extent to which individuals make use of certain services (such as family planning or abortion) in relation to distance from these services. While certain compositional factors will of course influence the take-up of these facilities, the influence of distance may also play a key role, as seen in a number of studies of family planning (Fuller 1984). Of the theories developed by other disciplines and applied in a spatial context, the most widely used - either explicitly or implicitly - has been that of modernisation theory and particularly in the way in which demographic transition theory has been applied in a spatial sense. In some ways the seemingly clear cut nature of modernisation theory is appealing in that fertility decline is associated with the process of economic development and socio-economic change and these relationships can be applied to variations between and within countries. In the latter context the model of convergence-divergence provides an interesting off-shoot of modernisation theory (Coward 1985), although the process of modernisation does not always spread with time or necessarily reduce regional economic differences in the long term. The advantage of modernisation theory is that it can be applied in a variety of settings and is sufficiently broad to incorporate many of the key variables affecting fertility. On the other hand, the disadvantages of such a framework - its rather rigid evolutionary assumptions, the diverse nature of the processes characterising modernisation and the ubiquituous collection of exceptions to the norm - ensure that modernisation theory cannot offer in itself an entirely satisfactory approach. Thus, depending on the scale of study, other theories may be of use when considering spatial variations in fertility, involving, for example, intergenerational wealth flows (Caldwell 1977, 1981, 1982), historical materialism (Woods 1982) and the new home economics (Schultz 1974). Furthermore, a more spatially explicit framework has been adopted by Eversley (1982), linking the spatial mobility of couples during their critical period of family building with their earning power, occupation and housing.

This framework was adopted to investigate regional demographies in Britain, but it is likely that such a framework can be applied in other settings also.

FERTILITY PATTERNS THROUGHOUT THE WORLD

Two general features characterise contemporary fertility patterns throughout the world: the considerable degree of variation and wide range of variables that influence such variations. Each of these themes are examined in this section.
While the considerable differences in fertility between those countries of the Northern and Southern blocs (as defined by the Brandt Report) are of quite long standing, more recent fertility declines in some countries of the South also ensure that there is now a considerable range of variation within this bloc. Thus while fertility remains high in some regions, most of Africa for example, more moderate levels have now been attained in much of South East Asia, Mid America and Tropical South America (Table 3.1). However, other areas have attained much lower fertility over the last three decades, as seen in the cases of Temperate South America, East Asia and the Caribbean (Table 3.1). Moreover, the differences between individual countries are particularly large: in Kenya, for example, currently displaying the highest level of fertility, average fertility is 8 children per woman (on the basis of current age specific rates), while in such countries as Cuba, Barbados, Singapore and Hong Kong the Total Fertility Rate is now 2 or less.

A variety of approaches can be utilised in attempting to explain these variations. In general terms it certainly appears that the relatively low fertility of some countries of the South is a product of various facets of modernisation and socio-economic development, associated with changing population composition and attitudes to family formation consistent with low fertility. Indeed, it is tempting to rely on the modernisation framework based on the demographic history of the current Westernised developed countries and assume that continued modernisation will in turn lead to low levels of fertility in the countries of the South. While it may appear unrealistic to expect a replication of demographic conditions amongst such a range of countries differing in terms of geographic and population size, political orientation, social structure and demographic history it has been argued that many parallels can be found (van De Walle and Knodel, 1980). However, using such a framework as modernisation does raise questions concerning whether general patterns can be discerned across the present developing countries and, if so, the specific variables involved. Analysis of World Fertility Survey data on differential fertility indicates that certain key factors can be identified (Lightbourne et al. 1982). From a study of 29 developing countries it was found that fertility

Table 3.1: Fertility Rates by Region for the Early 1980's

Region	Total population (millions)	Birth Rate	Total Fertility Rate	Per capita GNP (US $)
North Africa	124	41	5.9	1240
West Africa	161	49	6.8	660
East Africa	150	47	6.6	330
Middle Africa	60	45	6.2	420
Southern Africa	36	36	5.2	2490
S.W. Asia	110	37	5.5	4110
Middle South Asia	1036	37	5.3	260
S.E. Asia	393	33	4.5	720
E. Asia	1243	20	2.6	1360
Middle America	102	34	4.9	1970
Caribbean	31	26	3.4	-
Tropical South America	220	32	4.3	2120
Temperate South America	44	24	2.7	2440
N. America	262	15	1.8	13000
North & West Europe	236	12	1.7	11500
Eastern & Southern Europe	253	15	2.0	5270[1]
USSR	274	20	2.5	5940
Oceania	24	21	2.5	8700

[1]Southern Europe only.
Source: 1984 World Population Data Sheet (Population Reference Bureau).

is generally lower than average amongst those groups characterised by relatively high levels of female education and where females contribute to the non-domestic labour force in large proportions. Moreover, there is still a strong relationship between infant mortality and fertility (Lightbourne et al 1982). Boserup (1984) also emphasises the importance of the socio-economic factors previously mentioned, although in this case she broadens out the thesis by suggesting that one of the key factors influencing low fertility, is that of increasing security and status of women.

In some cases it is difficult to assess the precise effects of certain variables. For example, much discussion has centred on the role of family planning clinics in influencing fertility declines. Thus Cutright and Kelly (1981), analysing fertility change in 81 developing countries, suggest that this factor has the greatest impact on change while, on the other hand, Eberstadt (1980) argues that the importance of this factor has been exaggerated. Another perspective to declining fertility is offered by Caldwell (1977, 1981, 1982), arguing that fertility patterns must be considered in terms of family relationships in particular, suggesting that the main mechanism is that of the reversal of inter-generational wealth flows. He argues that the move away from family based production and the effects of Westernisation are of particular importance in bringing about these declines (Caldwell 1977, 1981, 1982). Another relevant factor is that of broader cultural and religious factors influencing fertility, as witnessed by the generally high fertility levels in many Moslem societies (Clarke 1984). Again it is difficult to assess the precise nature of the links because the general ideology of the Koran is not necessarily pro-natalist (IPPF 1982) and the key factor perhaps concerns the position of women in these societies, reflected in the more specific variables of relatively low levels of female education (compared with males) and limited non-domestic employment opportunities. Another important area of investigation concerns the role of political factors, particularly in the context by which governments perceive the level of fertility and the means by which such trends are subject to control and influence. Finally, it can be seen that some perspectives to fertility change run counter to the dominant themes concerning the nature of economic development, modernisation and demographic transition. Thus Harrison (1979), for example, suggests that in some agricultural settings the key factor is that of rural poverty, particularly in areas where there is great pressure on agricultural resources and with small average size of holdings. He argues that in some settings, particularly parts of rural Indonesia and Bangladesh, the population pressure on agricultural resources provides the motivation to lower fertility and thus, in some settings, there is a way out of the poverty trap. Harrison raises an interesting perspective although it should be emphasised that he provides little statistical evidence to support his hypothesis nor does he explain why these factors may be applicable in some settings but less relevant in other parts of the world which are likewise characterised by considerable pressure on agricultural resources, but where fertility remains high.

The role of certain of the more standard variables can be assessed for different groupings of the countries of the world using recently compiled data. In this case two groupings are used, representing all countries of the world for which adequate data exist and also those developing countries (defined as consisting of the Southern bloc). Fertility is measured by the Total Fertility Rate and thus provides an age standardised measure, although it can be noted that on both a worldwide and developing areas only basis there is a high degree of association between this measure and the birth rate ($r = 0.96$ and 0.94 respectively). The variables concerned and their hypothesised relationship with fertility are shown in Table 3.2 and while this type of study cannot assess the wide range of potential explanatory factors previously mentioned, it can, at least, indicate the nature of some of the broader aspects of certain factors linked to contemporary fertility patterns throughout the world. Several features are of interest. First, for example, at these scales of study the association between GNP and fertility is rather weak and is in fact not in the direction expected for the group of developing countries. In some ways this is surprising because the association tends to be stronger at coarser levels of aggregation, but in this case it may reflect the continuation of high fertility in some relatively prosperous parts of the developing world (such as the oil rich areas of SW Asia and North Africa) as well as the inability of average GNP to incorporate the important factor of income distribution. The more demographic variables of infant mortality and early marriage are quite strongly linked to variations in fertility and these factors are, in turn, quite strongly linked to some of the other socio-economic variables. Finally, while the association between fertility and female economic activity at this scale of study is very weak, there is a much higher degree of association between fertility and female education characteristics. Generally speaking most of these variables are highly intercorrelated and partial correlation analysis considerably reduces the degree of association with fertility (Table 3.2). However, the degree of association between fertility and infant mortality and also female literacy remains reasonably high (Table 3.2). Thus despite the problems of data availability and an inability to measure some of the major fertility influencing factors at this scale of study, it can be seen that certain broad demographic and socio-economic factors play important roles in differentiating contemporary fertility variation throughout the world.

Table 3.2: Relationships between the Total Fertility Rate and various demographic and socio-economic variables

Variable (and hypothesised relationship with fertility)		Zero Order Correlation/Partial Correlation			
		All Countries	Developing Countries	All Countries	Developing Countries
V1 GNP*	(-)	-0.24	0.04	-0.10	0.01
V2 Urban Population	(-)	-0.64	-0.47	0.03	0.03
V3 Labour force in agriculture	(+)	0.73	0.57	0.15	0.08
V4 Infant mortality	(+)	0.82	0.72	0.38	0.34
V5 Early marriage	(+)	0.69	0.59	-0.14	-0.15
V6 Females economically active	(-)	-0.13	0.03	-0.27	-0.20
V7 Female literacy	(-)	-0.87	-0.77	-0.28	-0.31
V8 Female Secondary school	(-)	-0.79	-0.67	-0.13	-0.20

* Logarithmic transformation

V2 % of total population living in areas termed urban by that country; V3 % of labour force engaged in agriculture; V4 infant deaths per 1000 births; V5 % females aged 15-19 in union; V6 females 15-64, % economically active; V7 % adults literate; V8 % 12-17 enrolled in Secondary School. Partial correlations refer to seventh order correlations between fertility and a specified variable controlling for the remaining seven variables.

Data source: 1984 world population data sheet; fertility and the status of women; 1982 world's children data sheet. All from the Population Reference Bureau.

The other feature of interest concerning fertility patterns that will be examined is that of the varying perceptions of governments towards fertility. Today, the potential for governments to influence fertility patterns is particularly great through the influence of the mass media, financial and socio-economic incentives and disincentives and the provision of family planning services. Thus the recent population policy in China directed towards the establishment of the one child family (Yan Tien 1983, Huang 1982) provides classic illustrations of how quickly fertility can decline in certain circumstances as well as the loss of individual freedom concerning some of the basic choices involved in family formation. Variations in governments' perception of fertility throughout the world are displayed in Table 3.3 and several features are of interest. Firstly, a large number of countries outside the North perceive fertility levels as being too high - these include all the countries with particularly large populations as well as many of the smaller ones too. In terms of population size, 89 per cent of the population of Asia reside in countries perceiving fertility as too high, while the corresponding proportions for Africa and Latin America are 51 and 30 per (Table 3.3). Many countries perceive fertility levels as satisfactory: most developed countries fall into this category but it also includes many of the countries and much of the population of Africa and Latin America. The precise reasons involved in these perceptions are not fully known, although it appears that many of these countries are not averse to reductions in fertility and in most cases some provision for family planning services is available. Finally numerous governments throughout the world perceive fertility as too low and hence may make concerted efforts to increase the birth rate (Table 3.3). As most attention is normally given to countries attempting to lower fertility, some characteristics of this final group will be assessed. Those countries perceiving fertility as too low are listed in Table 3.3 and although being dominated by many of the European countries, also includes examples from Africa, Asia and Latin America. The situation in some of the European countries is perhaps most readily understandable given the recent fertility declines to below replacement levels in many of these countries and the consequent concerns over population decline, shrinking labour force and ageing populations. France is an interesting example because the nature and implications of fertility trends have been a source of concern for much of this century. In the more recent period policy has been directed toward strengthening family policy provision whilst attempting not to appear explicitly pro-natalist (Huss 1980). A similar situation currently applies in West Germany (Kollman and Castell Rudenhausen 1982). In some Eastern European countries the governments have offered particularly attractive maternity benefits for women working outside the home in an attempt to encourage more three child families, as well as attempting to restrict the availability of abortion services. However, in areas such as the USSR and Hungary fertility has remained quite low and western researchers suggest that a shortage of suitable housing coupled with high costs of living may be the most important factors here (Freshbach 1982, Compton 1984). It is more difficult to assess the situation in those non-European countries because little information is available concerning the reasons behind such government perceptions and, in addition, fertility levels are generally high in these areas anyway. However, relevant factors might include a desire to increase population and the size of the armed forces either as a result of recent wars or conflicts (Israel and Kampuchia) or because of expansionist or the need to counter expansionist policies of others (Libya, Chad and Chile). Perhaps the most surprising examples are those in Middle and West Africa, where total fertility rates are around six children, but these perhaps reflect the strongly felt links between national status and population size in some African states.

FERTILITY PATTERNS IN THE BRITISH ISLES

Within the British Isles spatial variations in fertility are evident at a variety of scales and in this section some aspects of these patterns are examined within the separate constituent areas of England and Wales, Scotland, Northern Ireland and the Republic of Ireland. By way of background setting, it can be seen that there are certain broad differences between these areas: these generally reflect the contrast between the Republic of Ireland, with particularly high fertility in relation to most developed countries, and England and Wales, where fertility is much lower. These differences can be attributed to numerous cultural, socio-economic and demographic factors, fertility in the Irish Republic is, for example, between 50 and 100 per cent greater than that in England and Wales. Fertility rates in Scotland and Northern Ireland vary between these levels, with Scottish fertility very similar to that of England and Wales, while Northern Irish fertility occupies a position roughly halfway between that of England and Wales and the Republic of Ireland. Illegitimacy ratios, on the other hand, are lower in Northern and Southern Ireland. Current differences in nuptiality are generally less marked, although some of the features of the traditional West European marriage pattern (late age at marriage and low proportions marrying) still prevail in the Irish Republic. Some of the differences in nuptiality, particularly at young ages, can be attributed to the increasing proportion of couples cohabiting prior to, or instead of, marriage. In England and Wales the proportion of married females at ages 20-24 declined from 63 per cent in 1971 to 46 per cent in 1981, while somewhat smaller reductions were recorded for Scotland and Northern Ireland. Presum-

Table 3.3: Government perceptions of fertility level by area, number of countries and percentage distribution of population

Area	Government Perception of Fertility Level					
	Number of Countries			Percentage of Population		
	Too High	Satisfactory	Too Low*	Too High	Satisfactory	Too Low
Africa	26	19	6	51	45	4
Asia	17	18	4	89	10	1
Latin America	15	16	3	30	65	5
Oceania	4	3	0	22	78	0
Europe, USSR & North America	0	20	10	0	61	39

Source: 1984 World Population Data Sheet (Population Reference Bureau)

* The countries concerned are as follows:- France, Luxembourg, FRG, GDR, Sweden, Hungary, Romania, Bulgaria, Cyprus, Greece, USSR, Israel, Laos, Kampuchia, Libya, Chad, C.A.R, Congo, Gabon, Equatorial Guinea, Ivory Coast, Guinea, Uraguay, Chile, Bolivia.

ably, increasing rates of cohabitation, recorded for England and Wales by the General Household Survey (OPCS 1984a), have played a substantial role here. In the Irish Republic, on the other hand, nuptiality at these ages increased somewhat over the decade 1971-81.

In this section some characteristics of spatial variations in fertility will be examined in relation to England and Wales, Scotland, Northern Ireland and the Irish Republic. Two themes will be examined for England and Wales, involving the reduction in regional fertility differentials and recent variations in fertility at the local scale. For the remaining areas attention will be paid to recent spatial variations in fertility, with particular emphasis on differentials in marital fertility.

ENGLAND AND WALES: The declining magnitude of variation at the regional level

Within England and Wales the extent of variation in fertility at the regional level has generally declined over this century. Some characteristics of this decline are examined here, focusing on those factors which have played major contributory roles in this process. Fertility is examined in relation to the average family sizes of certain marriage cohorts measured from the Census Fertility Reports of 1951, 1961 and 1971. Although the data from these censuses are not identical in form, they are sufficiently similar to be utilised in comparing spatial variations in family size over the period 1926-1971. The average family sizes of the marriage cohorts 1926-31, 1936-41 and 1946-51, as measured from the censuses of 1951, 1961 and 1971 respectively, are selected for analysis, thus referring to the duration of marriage 20-24 years for each of the censuses and consequently representing completed or near completed fertility. For each cohort regional family sizes have been standardised for varying ages at marriage. The scale of study is fairly broad, representing the standard regions and with the major conurbations also included separately: in 1951 and 1961 this involves seventeen areas while in 1971 a slightly different set of units is utilised, as a result of boundary changes, identifying sixteen areas.

Some general features of the changing regional fertility of these cohorts are reflected in the changing coefficients of variation. The decline in the magnitude of variation is most noticeable when comparing the 1926-31 cohort (v = 9.8%) with that of the 1936-41 cohort (v = 6.0%) measured in 1951 and 1961 respectively, while there are only minimal differences between the cohorts measured in 1961 and 1971 (v = 5.8% in 1971). While the magnitude of variation declined from 1951 to 1961, the overall spatial patterning of fertility remained very much the same, as witnessed by the close association between the two distributions (r = 0.98). The spatial patterns of relative variation are shown in Figure 3.1 (where the overall figure for England and Wales for each of the cohorts is represented by 100) and again indicate that the variations are more marked in 1951. In 1951 the smallest average family sizes occur in London and the South East and the conurbations of West Yorkshire and South East Lancashire, contrasting with the larger average family sizes occurring in Merseyside, Central and Northern Wales, Tyneside and also the remainder of the Northern region (Figure 3.1). The main change between the cohorts measured in 1951 and 1961 was that the four areas with the highest fertility in 1951 had moved closer to the overall level but their fertility still remained generally higher than that of other areas - thus leading to the decline in the overall magnitude of variation but also maintaining the same broad spatial pattern. It is more difficult to assess precisely the changing distributions between 1961 and 1971 because of changes in regional sub-division over this period, by the general pattern in 1971 seems broadly comparable with the previous patterns. Throughout this period of study, the average size of family in the Merseyside conurbation has been markedly higher than that of any other area, a feature which can perhaps be attributed to the influx of relatively large numbers of migrants from Ireland coupled with the social class characteristics of the conurbation. Again it can be noted, however, that average family sizes have moved closer to the national level: in 1951 fertility was some 33 per cent higher, while the corresponding figure for 1971 was 18 per cent.

This decline in the extent of regional variation is a product of numerous processes but two would seem to have particular importance - the decline of social class differentials and the effects of regional convergence. Firstly, the period 1900-40 was one characterised by major declines in working class fertility as a result of a variety of social, economic, demographic and political factors (Gittens 1982) and, given the uneven spatial distribution of social class, this is likely to have had important repercussions on the spatial distribution of fertility. Unfortunately, data concerning family size by social class are not available for the separate areas and therefore this hypothesis remains difficult to verify, although the overall reductions in family size by social class for the marriage cohorts 1926-31 and 1946-51 are quite considerable, with the coefficient of variation dropping from 18 to 9 per cent (Table 3.4). For the marriage cohort measured in 1951 there is a fairly strong correspondence ($r = 0.65$) between regional average family sizes and social class (measured as the proportion of males in classes IV and V) and the narrowing of the overall class differential is therefore likely to have had important spatial consequences on fertility. The close association between the two variables was maintained in 1971 ($r = 0.84$). Another feature of interest is the disappearance of the purely inverse relationship between fertility and social

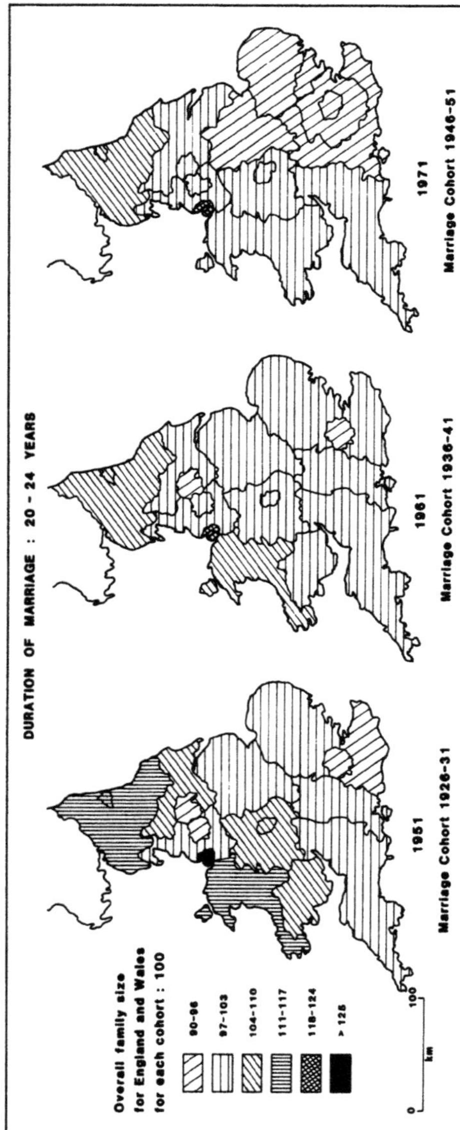

Fig. 3.1 Regional variations in family size in England and Wales: 1951, 1961 and 1971.

Table 3.4: Variations in family size by social class in 1951 and 1971, duration of marriage 20-24 years (standardised for age at marriage)

Social Class	1951		1971	
	Actual	Standardised	Actual	Standardised
I	1.77	1.86	2.13	2.18
II	1.91	1.98	2.03	2.06
IIIn	2.26	2.27	1.90	1.94
IIIm			2.25	2.17
IV	2.71	2.69	2.29	2.26
V	3.02	2.97	2.58	2.56
Coefficient of variation (%)	20.2	17.8	9.8	8.7

Source: Censuses of 1951 and 1971. Data refer to women married once only enumerated with husband. Data for 1951 refer to women aged less than 50 at time of Census, based on a 20% sample of the full Census.

class over this period (Table 3.4) which could also have blurred some of the more important differences in the regional distribution of family size.

The other important factor contributing to the convergence of regional fertility levels has been the regional convergence of certain key compositional factors influencing fertility. One such factor is that of female participation in the paid labour force and Gittens (1982), for example, argues that this factor was a major determinant of changing attitudes to family size by exposing women to varying norms and values concerning marriage and family formation, generating greater independence and by helping the development of joint role relations within marriage (a factor often associated with lower fertility). In 1951 there were moderate negative associations between regional family sizes and the proportions of females economically active ($r = -0.25$) and married women economically active ($r = -0.40$). Participation rates were higher than average in the West Yorkshire and South East Lancashire conurbations (dominated by textiles) and also Greater London (dominated by clerical positions), in contrast with the lower than average levels in more rural Central and North Wales and the Northern region (excluding Tyneside). However, in some instances the association between fertility and female participation is less clear cut, as seen in the cases of Merseyside (particularly high fertility and higher than average participation) and the South East (low fertility and participation). However, returning to the present hypothesis, a dominant feature of change between 1951 and 1971 was that the spatial distributions of participation became less concentrated. This is reflected in the changing coefficient of localisation between 1951 and 1971, which dropped from 30% to 9% for variations in married women economically active and from 17% to 8% for variations in all females economically active. These changes, reflecting the greatest relative increases in female participation in those areas with traditionally lower participation rates, could partly reflect the alterations in the spatial units over this period, although it seems unlikely that this could account for the major part of such changes. It can also be observed that the social class distributions of males also converged over this period, although not to the same extent, the coefficient of localisation dropping from 10% to 7%. Thus the period of study represents one of considerable convergence in terms of some of the compositional elements of the population and this, coupled with the wider dissemination of values and culture through the mass media, probably played an influential role in the declining extent of regional fertility variation.

Thus while this analysis has not been able to examine the range of variables studied by Gittins (1982), it seems likely that the declining differentials in fertility by social class and the trend toward regional convergence have played major roles in explaining changing regional fertility characteristics. Finally, while this section has concentrated on the declining extent of completed (or near completed) family size it should also be emphasised that regional variations in other aspects of fertility behaviour are still quite marked. For example, recent data concerning variations in illegitimate maternities, abortions and maternities legitimated following marriage (OPCS 1984b) indicate that the regional variations are still quite pronounced. Indeed, Overton (1982), in an analysis of fertility differentials in England and Wales, suggests that in some cases, such as the timing of first pregnancy or the control of pregnancies amongst the under twenties, the socio-economic differences may be widening. Thus regional convergence in one aspect of fertility may not necessarily indicate a general regional convergence in all fertility characteristics.

ENGLAND AND WALES: Variations at the district level

While contemporary regional variations in average family size in England and Wales are relatively small, there still exist quite pronounced variations in current fertility at more local levels. In many cases local age structures produce considerable variation in the birth rate, but, in addition, there are often marked variations in age standardised fertility measures. This section examines some characteristics of recent age standardised fertility in England and Wales with particular emphasis on patterns at the district level of aggregation.

The presence of considerable spatial variation in age standardised fertility at the district level is reflected in the fact that in 1981 approximately one quarter of all districts differed from the overall level for England and Wales by ten per cent or more. This compares with only 3 of the 53 areas at the county level. These districts with particularly high or low fertility are shown in Figure 3.2 and the immediate impression is that of a widely scattered distribution. There are, however, certain concentrations of such areas, such as in east Lancashire and parts of the Greater Manchester and West Yorkshire conurbations (high fertility); South Wales (high fertility); Inner London (generally low fertility); parts of the West Midlands conurbation, South Warwickshire and North Gloucestershire (low fertility) and the fringes of Greater London (generally low fertility). It is also possible to assess the contributions of each of the three main components of fertility - marital fertility, proportions married and extra-marital fertility - to these spatial patterns. The resulting classification tabulates areas of relatively high or low fertility in relation to the component (or combination of components) mainly responsible for influencing such patterns (Table

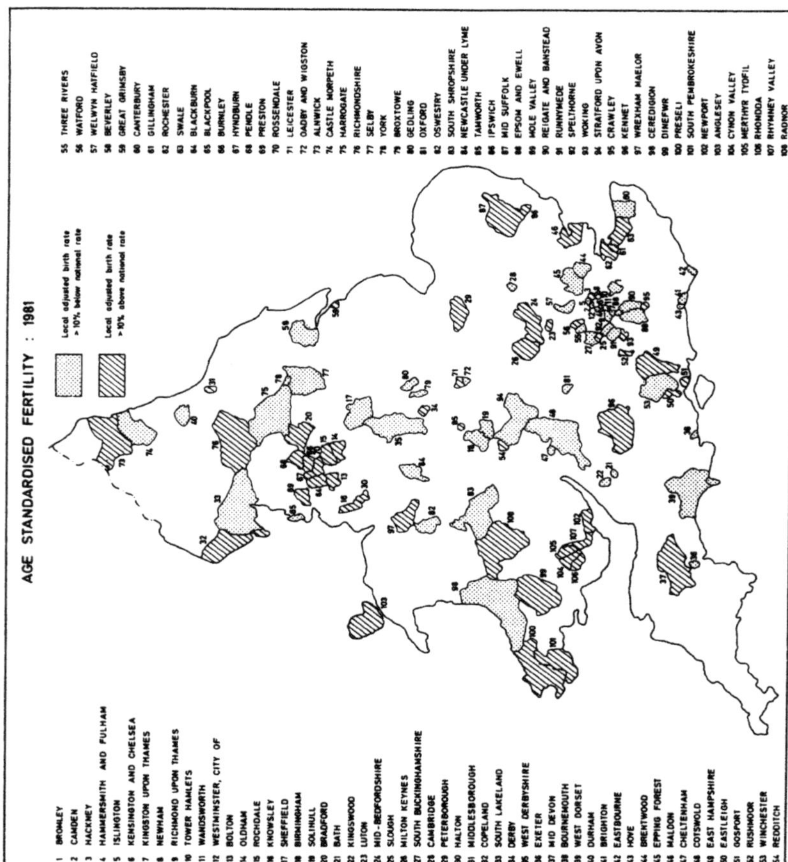

Fig. 3.2
Districts with particularly high or particularly low age standardised fertility, 1981.

3.5).[1] Generally speaking, the components for any one area tend to act in the same direction, such that areas with relatively high fertility tend to be associated with higher than average values for each of the components. However, various exceptions can be noted, the most notable being those areas in Inner London where low fertility is brought about by particularly low proportions married and occurs despite, in many cases, levels of marital and extra-marital fertility well above average.

Varying nuptiality occurs most frequently as the dominant factor characterising fertility, playing an important role, either singly or in combination with the other components, in 71 of the 108 cases. The corresponding figures for marital and extra-marital fertility are 41 and 37 respectively. The effect of relatively low nuptiality in bringing about low fertility is seen particularly for some areas in Inner London and certain medium sized towns in Southern England (Table 3.5). While these patterns may be influenced by relatively high levels of cohabitation in some areas, particularly Inner London, the major influencing factor would appear to be the migration of young single people to these areas, hence bringing about low nuptiality and low fertility. The fact that many of the towns and cities are service centres and university towns (such as Cambridge, Exeter, Canterbury and Oxford) is again likely to be a factor attracting migrants. Indeed, the high cost of housing in some of these towns may result in those couples in the early stages of family formation moving to outlying areas of cheaper housing. On the other hand, those areas of relatively high fertility influenced by high nuptiality are often small towns with new housing developments (such as Peterborough, Milton Keynes and Halton) which have attracted the migration of young couples in particular.

---

[1]For each area the expected numbers of legitimate, illegitimate and total births were derived from national age specific fertility rates. The difference between actual and expected total numbers was then apportioned into the relative excess or deficit of legitimate and illegitimate and the balance was assumed to reflect variations in proportions married. Single factor components were defined as those components where the excess or deficit of births associated with this factor were more than double the size of the next most important component. Double factor components were defined as those components each contributing at least 30 per cent to the excess or deficit while the third component contributed less than 20 per cent. Three factor combinations were defined as each component contributing at least 20 per cent to the excess or deficit of births.

Table 3.5: Districts with particularly high or low age standardised fertility in 1981 by major influencing factors

Single Component			Double Component			All three Components
Proportions married	Marital fertility	Extra-marital fertility	Marital fertility/ proportions married	Marital fertility/ extra-marital fertility	Proportions married/ extra-marital fertility	
A. Fertility > 10% higher than expected						
Mid. Bedford	Tower Hamlets	Middlesborough	Luton	Newham	Bolton	Bradford
Halton	Watford	Leicester	Slough	Birmingham	Burnley	Blackburn
Peterborough	Woking		Mid-Devon	Hackney	Rossendale	Pendle
Copeland	Crawley		East Hants	Knowsley	Ipswich	
Maldon			Alnwick	Derby	S. Pembroke	
Eastleigh			Wrexham	Gt. Grimsby	Oldham	
Gosport			Anglesea	Preston	Rochdale	
Rushmoor			Radnor		Newport	
Redditch						
Rochester						
Hyndburn						
Richmondshire						
Tamworth						
Mid-Suffolk						
Kennet						
Dinefwr						
Preseli						
Merthyr Tid.						
Rhymney						
Milton Keynes						
Suble						
Gillingham						
Rhondda						

B. Fertility > 10% lower than expected

Bromley	Sheffield	S. Lakeland	Bournemouth	W. Derby	S. Bucks
Camden	Solihull	Epping Forest	Cheltenham	W. Dorset	Brentwood
Hammersmith	Kingswood	Winchester	Blackpool	Durham	Welwyn Hatfield
and Fulham	Broxtowe	Harrogate	S. Shrops	Cotswold	Epsom
Kensington	Gedling			Beverley	Reigate
and Chelsea	Newcastle U.L.			Oadby	Runnymede
Richmond	Oswestry			Castle Morpeth	Spelthorne
Wandsworth				Selby	
Westminster				Stratford U.A.	
Bath					
Cambridge					
Exeter					
Brighton					
Eastbourne					
Hove					
York					
Three Rivers					
Canterbury					
Oxford					
Ceredigion					

See Footnote [1] for explanation of terms.

Variations in marital fertility also play an important role in characterising areas of relatively high and low fertility. Thus in four cases it plays a dominant role in influencing high fertility (Table 3.5) and in three of these cases the areas concerned are towns characterised by the development of housing schemes over the last fifteen years (Watford, Woking and Crawley). This feature seems consistent with Eversley's (1982) thesis that many young couples intending family building move to areas with reasonably good access to cheaper housing in order to use their financial resources most efficiently at a time when such resources are often stretched. This feature also seems applicable to the cases of Luton and Slough, where relatively high marital fertility and proportions married contribute to high fertility. In other areas the high marital fertility may be a result of the relatively large proportions of the population of New Commonwealth and Pakistan origin, as seen, for example, in Tower Hamlets and, combined with high extra-marital fertility, Newham and Hackney (Inner London) and also Birmingham. On the other hand, particularly low marital fertility brings about low fertility in a variety of different areas (Table 3.5). In some cases, for example, the relatively high social status of the areas concerned may lead to lower fertility as in Solihull and Kingswood and, combining with one of the other components, West Derbyshire, Beverley and South Shropshire.

The contribution of extra-marital fertility in influencing age standardised fertility is often in conjunction with one of the other components. Here it is more difficult to assess the spatial patterns because variations in extra-marital fertility by socio economic grouping are not fully documented although it is known that rates are generally high amongst immigrant groups of Caribbean origin. More recently, the link between high extra-marital fertility and high youth unemployment has been suggested (Campbell 1984). On this basis, some areas with relatively high fertility influenced wholly or partly by high extra-marital fertility can be linked with relatively large Caribbean immigrant populations (Hackney, Birmingham and Newham) or high levels of male and female youth unemployment (Middlesborough, Knowsley and South Pembrokeshire). Many of the areas with low fertility in which low extra-marital fertility plays a contributory role tend to be associated with relatively high social status and low levels of youth unemployment, as seen in the cases of Winchester, Harrogate and South Buckinghamshire.

It can also be noted that the spatial variations around London seem to follow a distinct pattern in that while Inner London is associated with areas of both relatively high and low fertility, many of the areas of the suburban fringe of Greater London are associated with low fertility (particularly on the northern, western and south western edges) while, on the other hand, many of the outlying areas are associated with high fertility (such as Luton, Milton Keynes, Gillingham, Rochester and Rushmoor). These latter features can again perhaps be explained in relation to Eversley's (1982) thesis that couples in the early phase of family formation (those with high level of current fertility) may be inclined to move to more outlying areas in search of cheaper housing, while couples who have delayed or finished family formation may have greater disposable resources for the generally higher house prices in the outer suburbs of Greater London. It is difficult to verify this hypothesis from the evidence presented here, but the regularity of the spatial patterns in and around London would suggest that certain key socio-economic factors are operating.

Finally, the overall variations in age standardised fertility for the total number of districts (403) in England and Wales can be assessed in relation to a range of standard demographic and socio-economic variables derived from the census. Some of these factors prove difficult to measure as in, for example, the effects of the migration of young couples where, due to unavailability of more precise information, the rate of net migration over the period 1971-81 for the whole population is utilised. The variables concerned and their relationship with fertility are shown in Table 3.6. The zero order correlations are all in the expected direction, with four of the eight greater than 0.35. Five of the variables make a significant contribution to the overall variations in fertility in the context of a stepwise regression model, accounting for 50 per cent of the variation (Table 3.6). Apart from marital status, the most important socio-economic variables are those of female unemployment, immigrant populations, female education and the participation of married women in the labour force. However, the latter two variables only contribute marginal increases in the proportion of the variation explained. Thus it can be seen that on an overall basis certain standard demographic and socio-economic variables account for about one half of the variation in age standardised fertility in England and Wales, indicating that the search for further relevant variables should continue. Future analysis, for example, will hopefully utilise a more appropriate measure of migration than the one utilised here as well as examining housing characteristics in more detail.

FERTILITY VARIATIONS IN SCOTLAND

The pioneering contributions to the study of recent variations in Scottish fertility have been those of Jones (1975) and Wilson (1978) who have examined variations in the birth rate in 1971-72 and variations in family size measured in 1971 respectively. Both conclude that there are quite marked differences in fertility at the county level of aggregation and

Table 3.6: Relationship between age standardised fertility and various demographic and socio-economic variables, 1981

	Variable (and hypothesised relationship with fertility)	Zero order correlation with fertility	% Contribution to variation (cumulative % from stepwise regression)
V1	Proportions married (+)	0.54	29
V2	Female Unemployment (+)	0.37	40
V3	Immigrant Population* (+)	0.01	48
V4	Female Education (-)	-0.46	49
V5	Participation of married women (-)	-0.13	50
V6	Social classes (+)	0.39	-
V7	Tenure: owner occupation (-)	-0.06	-
V8	Migration (+)	0.10	-

Regression Equation:-

$$Y = 57.4 + 0.01\ V1 + 0.11\ V2 + 9.0\ V3 - 0.04\ V4 - 0.02\ V5$$

* Logarithmic transformation.
V1 % females 15-29 married; V2 proportion of females aged 16-59 unemployed; V3 % of population born in the New Commonwealth, Pakistan and Ireland; V4 % of females with higher education; V5 % of married women aged 16-59 economically active; V6 % of population with heads of household in social classes IV and V; V7 % of private households owner occupied; V8 overall net migration rate 1971-81.

that certain key variables play an important role in contributing to these variations. This section examines fertility patterns in 1981 and attempts to discern if the characteristics observed in 1971 are applicable to more recent variations. The study concentrates on current fertility because questions relating to children ever born were omitted from the United Kingdom census of 1981.

Jones' (1975) study emphasises that variations in age and marital structure play a major role in influencing spatial variations in the birth rate in 1971-72. On this basis, therefore, it might be expected that the extent of variation in those fertility measures taking account of such variables would be much less than that of the birth rate. However, this does not appear to be the case and for 1981 the magnitude of relative variation in the birth rate (coefficient of variation = 11%) is only slightly greater than that for the index of overall fertility (10%), which indirectly standardises for age, and the index of marital fertility (9%), accounting for age and marital structure. Interestingly the extent of relative variation in the illegitimacy rate (30%) is particularly large. The relative degree of variation in the birth rate in 1981 was somewhat higher than that for 1971, although the current study is based on a larger number of spatial units as a result of boundary changes (56 as opposed to 31 used by Jones) and consequently provides a more detailed summary than that available for 1971.

In 1981 the general pattern of variation in the birth rate was broadly similar to that of 1971 (allowing for differences in the spatial units utilised), with relatively low rates amongst the border areas, the Tayside region and the cities of Aberdeen, Dundee and Edinburgh. This contrasts with the generally higher rates in much of urban central Scotland, particularly the eastern side of the Glasgow conurbation. One difference, however, is that in 1981 relatively high birth rates occur in many of the counties of the Highlands and Islands, whereas this feature was much less apparent in 1971-72. Variations in age standardised fertility can be best illustrated by using Coale's indices, in which variations in overall fertility reflect the varied contributions of marital fertility, proportions married and extra-marital fertility.[2] Indeed, variations in these components illustrate that the spatial patterns are by no means similar (Figure 3.3). For example, the pattern of variation in marital fertility is quite different from that of variations in either proportions married

[2]Coales indices are age standardised measures based on a comparison between actual and expected births, where the latter are derived from the Hutterite age specific fertility rates 1921-30 which are taken to represent natural fertility. See Coale (1967).

or extra-marital fertility (r = -0.11 and 0.15 respectively), with the spatial pattern of marital fertility dominated by the relatively high rates of many of the west coast highland areas, the Western Isles, Moray and some eastern areas of the Glasgow conurbation (Figure 3.3). Indeed this pattern is broadly similar to that of variations in family size as examined by Wilson (1978), emphasising the continuity in marital fertility over time and across both cross-sectional and cohort measures. Both Jones and Wilson show that certain key compositional variables such as female activity, social class and religion play a significant part in accounting for variations in fertility and it would thus be expected that variations in marital fertility in 1981 would also be partly explained by such variables. The strength of these associations are summarised in Table 3.7 and the variables concerning the proportion of the labour force in agriculture and housing tenure have also been added to the analysis In the case of the latter, it has been shown recently that tenure can play a major role in contributing to differential fertility in Britain (Murphy and Sullivan, 1983). The correlations between marital fertility and the remaining variables are all, with the exception of tenure, in the hypothesised direction and the variable with the greatest explanatory power is, similar to Wilson's study, that of the participation of married females in the paid labour force (Table 3.7). In a stepwise regression model, two of these compositional variables are significant (participation of married women in the paid labour force and religious denomination), accounting for 60 per cent of the variation in marital fertility in 1981. On the other hand, these compositional variables play a much less important role in accounting for variations in proportions married, where the spatial patterns are less clearly discernible. In this case low nuptiality occurs in the major cities (perhaps due to the in-migration of young single females to these areas as well as a greater degree of cohabitation), the western isles and parts of the Highland and Grampian regions. Interestingly, the pattern of illegitimacy is also varied, reflecting the high rates in the urban areas of Glasgow and Dundee and some areas in the Highlands such as Ross and Cromarty and also Sutherland (Figure 3.3).

Broad patterns of illegitimacy in Scotland have changed considerably over the last fifty years from one where, during the pre-war period, many rural areas tended to have relatively high levels to one where, since the mid 1950's, large urban areas have generally had the highest levels (Illsley and Gill, 1968). The current distribution, therefore, seems to reflect elements of both the old and the new patterns. The index of overall fertility is influenced by each of the three constituent components of fertility, but in this case it appears that spatial variations in marital fertility play the dominant role in influencing overall fertility. For example, the degree

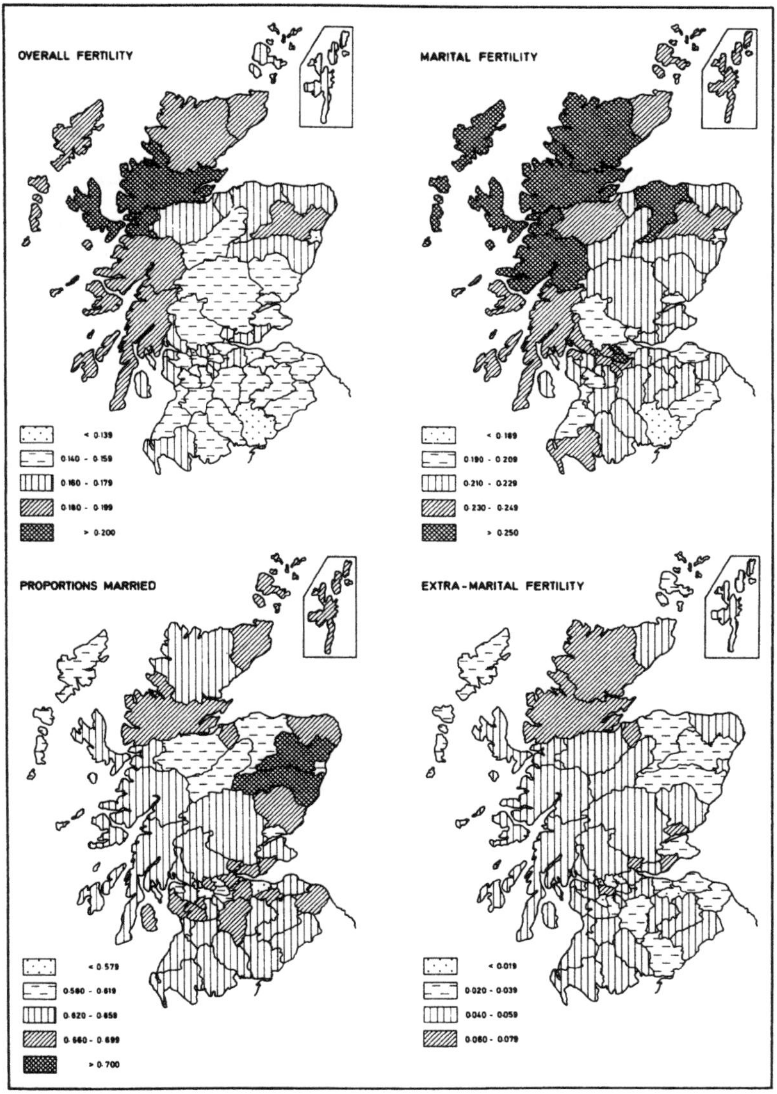

Fig. 3.3 Variations in Coale's indices of fertility, Scotland 1981.

of association between these two indices (r = 0.84) is much higher than the degree of association between overall fertility and each of the other two components. Thus overall fertility is higher in the western coast counties, the Western Isles and the northernmost coast counties with lower rates generally occurring throughout the rest of the country (Figure 3.3). Of the previously mentioned socio-economic factors, only that of the participation of married women in the labour force is highly associated with overall fertility, in this case accounting for 59 per cent of the variation.

To conclude, while it is difficult to compare the spatial patterns in 1981 with those of 1971 due to changing boundaries and varying data availability, there appear to be certain consistent features concerning recent Scottish fertility patterns. Thus current variations are still quite marked and are by no means a product of varying age and marital structure alone. Indeed certain compositional variables play a key role in influencing the spatial patterns and while the exact nature of the causal relationships cannot be precisely specified from the available aggregate data, it appears that variations in overall and marital fertility at this scale of study can be partly attributed to differences in female activity patterns, while variations in religious denomination exert an important influence on varying marital fertility.

DIFFERENTIALS IN THE IRISH REPUBLIC

There are several ways in which fertility patterns in the Republic of Ireland are of interest. Firstly, for example, the overall level of fertility is particularly high by West European standards and this feature can be attributed to a variety of economic, social, cultural and demographic factors (Coward 1978, 1980; Kennedy 1973). Secondly, fertility is in a state of rapid change, involving a trend towards earlier and more universal marriage coupled with greater control of fertility within marriage. The pace of these changes has been particularly marked since 1971, indicating that the population of the Irish Republic is rapidly moving away from the Malthusian form of population limitation (delayed marriage and relatively high celibacy coupled with little restriction of fertility within marriage) towards more neo-Malthusian forms characteristic of most European neighbours (Coward 1982). However, while fertility has generally declined over the last fifteen years (the birth rate dropping from 22 per thousand to 19 per thousand between 1971 and 1983), it still remains relatively high in a West European context. Thirdly, spatial variations in fertility are quite marked, even at fairly coarse levels of aggregation, tending to reflect, in particular, differences in economic development and occupation structure (Coward 1978, 1980). This section briefly considers some of the salient

Table 3.7: Relationships between marital fertility and hypothesised explanatory variables, Scotland 1981

Variable (expected relationship with marital fertility in parentheses)		Zero Order Correlations V1	V2	V3	V4	V5	V6
V1	Marital fertility	1.00	0.23	-0.34	-0.69	0.14	0.07
V2	Religion (+)		1.00	0.19	0.07	-0.41	-0.60
V3	Social class (-)			1.00	0.50	0.20	-0.50
V4	Female participation (-)				1.00	-0.09	-0.38
V5	Owner-occupation (-)					1.00	0.31
V6	Labour force in agriculture (+)						1.00

V2: Proportion of marriages celebrated in the Roman Catholic Church.
V3: Proportion of married women whose husband's social class is skilled non-manual (IIIn).
V4: Proportion of married women aged 16-44 in paid employment.
V5: Proportion of private households owner occupied.
V6: Proportion of males in agriculture.

characteristics of recent spatial variations in fertility at the county scale of aggregation and the nature of the changing spatial patterns over the inter-censal decade 1971-1981.

As for many countries, variations in the Irish birth rate are strongly influenced by age and marital structure. This is observed, for example, when contrasting the relatively low rate (16-20 per thousand) in the more economically depressed north-western parts of the country (traditionally associated with elderly age structures and low nuptiality) with the generally higher rates (23-27 per thousand) of the eastern counties around Dublin. Indeed, these two variables (measured by the proportions of the population aged 20-34 and females married at ages 15-49) account for 71 per cent of the variation in 1981. The particularly high rates of the eastern counties around Dublin - in excess of 23 per thousand - ensure that natural increase is particularly high in these areas and, coupled with high rates of in-migration, place considerable pressure on the availability of land, housing and resources in these areas. However, while age structure has a major influence on the birth rate, it can be seen that considerable variations in fertility are still present after controlling for this variable - as indicated by the overall degree of variation (v = 12%) in the Total Fertility Rate compared with the birth rate (v = 11%).

Changing patterns of age standardised fertility over the decade 1971-81 can be best illustrated by utilising Coale's indices in which the separate contributions to overall fertility by marital fertility and proportions married (illegitimacy rates are generally low outside Dublin) can be assessed. Indeed, the particularly marked changes in fertility which have occurred during the decade 1971-81 have involved contrasting trends in each of the main components of fertility. Thus, on a national basis, proportions married increased by 8 per cent while marital fertility declined by 25 per cent and the particularly large decline in the latter was responsible for the general decline of 18 per cent in the overall level of age standardised fertility (Figure 3.4). Generally speaking these changes took place throughout the counties and county boroughs of the Irish Republic such that the population in all of these areas underwent declines in overall fertility and marital fertility, while the population of all but three areas (those urban populations of Dublin, Cork and Dun Laoghaire particularly affected by continued in-migration of young single people) underwent increases in proportions married (Figure 3.4). Interestingly, the greatest relative increases in proportions married occurred in those western counties which have generally displayed lower than average nuptiality and therefore it appears that the broad regional patterns of nuptiality are converging. The decline in marital fertility is of particular interest for two reasons. First, the general decline of 25 per cent is a particularly large reduction over the

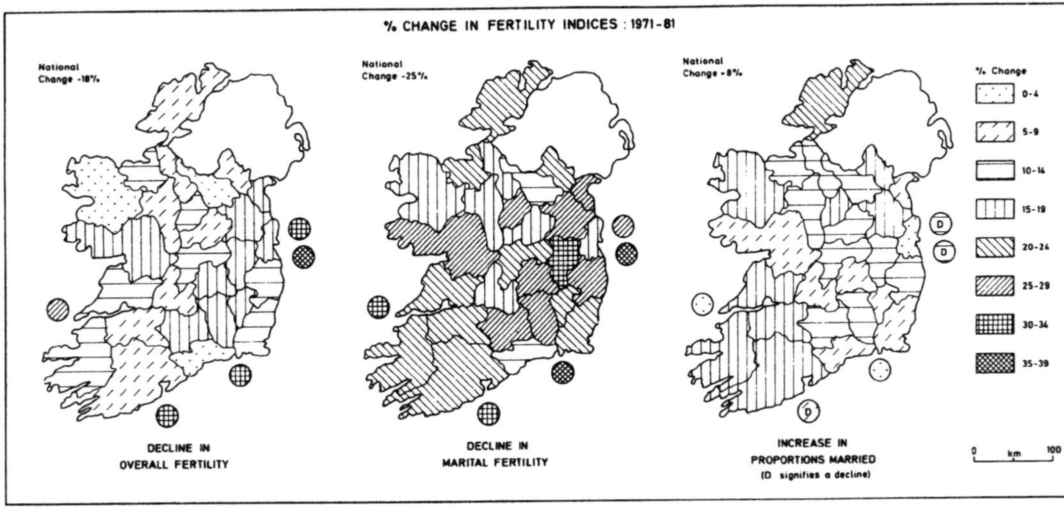

Fig. 3.4 Changes in the indices of fertility 1971-81 in the Republic of Ireland. (The 5 urban boroughs are circled).

decade and is surely a product of the various social, economic, cultural and demographic changes that occurred during the 1970's. Of these the most important regarding fertility are the continued declines in agriculture, the growth of an urban based middle class, return migration from Britain, increasing employment opportunities for married women, the growth of the family planning movement, increasing strength of feminism and rising secularisation amongst the younger sections of the population in particular. The second point of interest is that the declines in marital fertility have been quite large throughout the whole country and were by no means confined to the principle urban areas (Figure 3.4). This suggests that the scale of social change was also widely spread and in this context the movement of return migrants to many of the more rural parts of Ireland coupled with government policy aimed at fostering economic development and industrial growth throughout the whole country seem particularly appropriate in attempting to understand the changing attitudes to family formation and fertility. The declines in overall fertility have been particularly marked in the main urban areas (due to large reductions in marital fertility coupled with small or no increases in proportions married), while the reductions amongst the populations of the eastern counties have generally been greater than those of the western counties. However the fact that overall fertility has declined throughout the whole country (at this scale of study) remains the dominant characteristic of changing fertility patterns over the most recent inter-censal decade.

FAMILY SIZE IN NORTHERN IRELAND

Similar to the Irish Republic, marked spatial variations in fertility are evident in Northern Ireland. Compton (1978a, 1978b), for example, has examined the detailed patterns around 1971 and shows that the distribution of the population by religious denomination, along with socio-economic and demographic factors, play important roles in accounting for these variations. While questions relating to family size were omitted from the 1981 census, a fertility survey undertaken throughout Northern Ireland in 1983, based on a sample of 3000 ever-married women and their partners, allows more recent trends to be investigated. Some of the preliminary results of the survey are recorded in Table 3.8 and in this case five separate areas are given in relation to average completed family size. It is apparent that the differences in family size between Roman Catholics and non-Roman Catholics are still very distinctive and for both of these groupings there is a quite marked spatial gradient of fertility. For example, average family sizes are generally smallest amongst the population living beyond the immediate Belfast suburbs,

Table 3.8: Variations in family size in Northern Ireland by area and religion, 1983. (Ever-married women aged 45-59)

Area	Roman Catholic	Non-Roman Catholic
Belfast	4.0	2.4
Belfast Suburbs	3.7	2.6
Inner Zone: Fringe Belfast	3.3	2.2
Intermediate Zone	5.2	2.7
Outer Zone: West of the Bann	5.4	3.2

Source: Northern Ireland Fertility Survey, 1983.

somewhat greater in the intermediate fringe areas and particularly large amongst the more rural populations in the western half of the province. Further analysis will be able to confirm the extent to which this gradient is a product of certain compositional variables such as social class, education, housing and female activity, or whether it is influenced by other factors such as the diffusion of the small family ideal from the more developed east to the less developed west. The example of Northern Ireland therefore provides a suitable illustration of marked variations in fertility in (relatively small) developed areas and that these are evident even at fairly coarse levels of aggregation.

CONCLUSION

The chapter has emphasised some of the characteristics of contemporary fertility patterns for a variety of different scale levels and from the examples cited here it appears that variations in fertility are most evident towards the extremes - both micro and macro - of the scale continuum. Given the relative neglect of fertility studies by geographers, future research will hopefully contribute to the continued documentation and monitoring of spatial fertility patterns throughout the world at a variety of scales, as well as contributing to methodological developments by, for example, examining a wider range of fertility influencing variables, making greater use of individual data as well as assessing the role of general models of fertility change and decline.

REFERENCES

Boserup, E. (1984) 'Shifts in the Determinants of Fertility in the Developing World: Environmental, Technical, Economic and Cultural Factors,' Paper presented at B.S.P.S. Annual Conference, Cambridge

Caldwell, J. (1977) 'The Economic Rationality of High Fertility: An Investigation Illustrated with Nigerian Survey Data', Population Studies, 31, 5-27

Caldwell, J. (1981) 'The Mechanisms of Demographic Change in Historical Perspective', Population Studies, 35, 5-27

Caldwell, J (1982) 'The Failure of Theories of Social and Economic Change to Explain Demographic Change', Research in Population Economics, 4, 297-332

Campbell, B. (1984) Wigan Pier Revisited: Poverty and Politics in the 1980's, Virago Press, London

Clarke, J. (1977) 'Population Geography', Progress in Human Geography, 1, 136-41

Clarke, J. (1984) 'Islamic Populations: A Synthesis,' Paper presented at I.B.G. Annual Conference, Durham

Coale, A. (1967) 'Factors Associated with the Development of Low Fertility: An Historic Summary', World Population Conference 1965, 2, 205-209

Compton, P. (1978a) 'Fertility Differentials and their Impact on Population Distribution and Composition in Northern Ireland', Environment and Planning, A, 10, 1397-1411

Compton, P. (1978b) Northern Ireland: A Census Atlas, Gill and Macmillan, Dublin

Coward, J. (1978) 'Changes in the Pattern of Fertility in the Republic of Ireland', Tijd. voor Econ. en Soc. Geog. 69, 353-61

Coward, J. (1980) 'Variations in Family Size in the Republic of Ireland', Journal of Biosocial Science, 12 (1), 1-14

Coward, J. (1982) 'Fertility Changes in the Republic of Ireland During the 1970's', Area 14 (2), 109-117

Coward, J. (1985) 'The Spatial Perspective in the Study of Fertility and Mortality', in R. Woods and P. Rees (eds.), Developments in Spatial Demography, George Allen and Unwin, London

Cutright, P. and Kelly, W. (1981) 'The Role of Family Planning Programs in Fertility Declines in Less Developed Countries', 1958-1977, International Family Planning Perspectives 7, 4, 145-151

Eberstadt, N. (1980) 'Recent Declines in Fertility in Less Developed Countries, and What "Population Planners" May Learn From Them', World Development 8, 37-60

Eversley, D. (1982) 'Demographic Change and Regional Policy in the United Kingdom', in D. Eversley and W. Kollman (eds.), Population Change and Social Planning, 349-373, Edward Arnold, London

Freshbach, M. (1982) 'The Soviet Union: Population Trends and Dilemmas', Population Bulletin 37, 2, Population Reference Bureau

Fuller, G. (1984) 'Population Geography and Family Planning', in J. Clarke (ed.), Geography and Population, Pergamon, Oxford
Gittins, D. (1982) Fair Sex: Family Size and Structure, 1900-39, Hutchinson, London
Harrison, P. (1979) 'Poverty and Population', New Society, 5 July, 9-11
Huang, L. (1982) 'Planned Fertility of One Couple-One Child Policy in the People's Republic of China', Journal of Marriage and the Family, 775-784
Huss, M. (1980) 'Demography, Public Opinion and Politics in France 1974-80, Occasional Paper No. 16, Dept. of Geography, Queen Mary College, London
Illsley, R. and Gill, D. (1968) 'Changing Trends in Illegitimacy', Soc. Sci. and Med. 2, 415-33
International Planned Parenthood Federation (1982) 'Islam and Family Planning Leaflet 1/82, IPPF, London
Jagielski, A. (1980) 'Urbanisation and Spatial Aspects of "Demographic Transition" Oeconomica Polona, 1, 84-104
Jones, H. (1975) 'A Spatial Analysis of Human Fertility in Scotland, Scot. Geog. Mag, 91, 102-113
Jones, H. (1981) A Population Geography, Harper and Row, London
Jones, H. (1984) 'Population Geography in Britain', in J. Clarke (ed.), Geography and Population, Pergamon, Oxford
Kennedy, R. (1973) The Irish: Marriage Fertility and Emigration, University of California Press, London
Kollman, W. and Castell Rudenhausen, A. (1982) 'Past and Present Policy Reactions to Fertility Decline in Germany', in D. Eversley and W. Kollman (eds.), Population Change and Social Planning, Edward Arnold, London, pp. 414-24
Lightbourne, R. Singh, S. Green, C. (1982) 'The World Fertility Survey: Charting Global Childbearing', Population Bulletin 37, 1, Population Reference Bureau
Murphy, M. and Sullivan, O. (1983) 'Housing Tenure in Fertility in Post-war Britain', Centre for Population Studies Research Paper No. 83-2
OPCS (1984a) General Household Survey, 1982 HMSO, London
OPCS (1984b) 'OPCS Monitor: Conceptions Inside and Outside Marriage, 1969 to 1981', Reference FMI 84/86, OPCS, London
Overton, E. (1982) 'The Decline in Fertility Since 1964', in D. Eversley and W. Kollman (eds.), Population Change and Social Planning, Edward Arnold, London, pp. 20-61
Population Reference Bureau (1984) '1984 World Population Data Sheet', P.R.B., Washington D.C.
Rees, P. (1981) 'Population Geography', in N. Wrigley and R. Bennett (eds.), Quantitative Geography: a British View, Routledge and Kegan Paul, London

Schnell, G. and Monmonier, M. (1983) The Study of Population: Elements, Patterns and Processes, Merrill, Columbus, Ohio
Schultz, T. (ed.), 1974 Economies of the Family: Marriage, Children and Human Capital, University of Chicago Press, Chicago
Trewartha, G. (1953) 'A Case for Population Geography', An.Ass.Am.Geogs. 43, 71-97
Van de Walle, E. and Knodel, J. (1980) 'Europe's Fertility Transition: New Evidence and Lessons for Today's Developing World', Population Bulletin, 34, 5, Population Reference Bureau
Wilson, M. (1978) 'A Spatial Analysis of Human Fertility in Scotland; Re-Appraisal and Extension', Scot. Geog.Mag. 94, 130-143
Woods, R. (1979) Population Analysis in Geography, Longman, London
Woods, R. (1982) Theoretical Population Geography, Longman, London
Yuan Tien, H. (1983) 'China: Demographic Billionaire', Population Bulletin 38, 2, Population Reference Bureau

Chapter Four

MORTALITY PATTERNS IN THE MODERN WORLD

P. H. Curson

GEOGRAPHY AND MORTALITY

Population Geography - Mortality Mislaid

The analysis of mortality remains the cinderella subject of population geography. Largely, population geographers have turned their back on mortality studies and few today include it as one of their major interests. Most geographers interested in the study of mortality have either moved laterally into medical geography or, like the author, have taken refuge within historical demography or population studies. It is consequently somewhat difficult to address the subject of mortality patterns in the modern world in a volume designed to discuss progress in population geography when very little work on mortality has been carried out by population geographers. Rather, the spatial study of mortality remains a central concern of medical geography and, while a strong theoretical case can be made for regarding medical geography as an integral part of population geography, the current status enjoyed by medical geography is clearly that of a viable and active sub-discipline in its own right (Jones, 1984, 173). The paucity of studies of mortality by population geographers is partly due to their preoccupation with the more traditional geographical aspects, such as the relationships between population distribution and environment factors or between total population numbers and human resources. Specifically there has been a concentration on the description and analysis of country, regional and international population patterns, usually in terms of the totality of population features with general explanations sought from the broadest aggregative data. Only very rarely have population geographers concerned themselves with the systematic study of particular characteristics such as fertility and mortality, and rarely has there been any concerted effort to investigate well-defined causal hypotheses. Almost all the texts on the subject to appear before 1970 were heavily pattern-oriented in approach, giving extensive coverage to the spatial distribution of population, including demographic, socio-economic and ethnic factors, migration, and the population-resources question, but with very little attention to fertility and mortality per se. From the 1970s, despite a general swing towards a more rigorous process-oriented approach in geography generally, and the eloquent counsel of some practitioners such as Woods (1979) and Jones (1981) for a more single-minded pursuit of causal explanations particularly in the study of fertility and mortality, population geographers have continued to be occupied with 'matters of distribution, with identifying population types and especially with migration...' (Woods, 1984, 44).

The lack of attention given to mortality also stems from the way in which geography in general and population geography in particular have developed in the post-1960 period. Increasing specialisation and fragmentation have seen the emergence of viable sub-areas of specialisation such as medical geography and urban social geography, so that Clarke could suggest, even as early as 1972, that many population geographers excluded any detailed micro-analysis of mortality patterns from their range of interests for fear of encroaching upon the domain of medical geography (Clarke, 1972, 126). Given such attitudes, the emergence of historical demography, with its concern for reconstructing past mortality patterns, and epidemiology, with its broad interest in mortality and disease distributions, has further eroded a substantial part of population geography's traditional field of interest. Only in recent years, mainly through the work of Woods and Jones, has there been any clarion call for a recognition of the role population geography can play in the analysis of mortality patterns. Generally, however, such a call has gone unheeded, and by the 1980s population geography had for many come to represent the spatial study of migration, population redistribution, and broad aggregative analyses of spatial distributions. In 1984 Clarke, the Chairman of the International Geographical Union's Commission on Population Geography, seems to have accepted the fact that mortality (and fertility and population characteristics) no longer comprised a significant part of population geography. Speculating on the direction the discipline might take in the next few years, he argued that migration (particularly population redistribution, labour movement, outmigration, transmigration, and refugee movement), decentralisation, urbanisation and disasters could represent the areas of future concern (Clarke, 1984, 8-9).

Medical Geography - Mortality Retrieved

The end result of the above was that, by the 1970s, most of the substantive work on mortality within geography was being undertaken by medical geographers and historical demographers. In many cases such work has pursued with great vigour and evangelical zeal the broad theme of observing, documenting, and analysing spatial variations in human mortality and health patterns and their environmental relationships. Broadly, medical geographers have claimed that the identification of the spatial contours of mortality and its physical, social and economic correlates at varying levels of scale may ultimately be productive of hypotheses leading to causal explanations. Some of the problems associated with this claim and with the geographic approach to mortality will be considered later in this chapter, but for the moment it may be apposite to mention just one. One of the ironies of medical geography is that its primary interest in the occurrence, distribution and determinants of health in human populations has often been pursued through the analysis of mortality patterns. In the absence of detailed records of morbidity, medical geographers have often been forced to rely on spatial patterns of mortality rates by specific cause as a surrogate indicator of geographic variations in ill-health. The problem is, of course, that death rates may reveal relatively little about illness patterns in a population, and perhaps even less about the role of medical intervention. The question remains whether the residential distribution of people who died from ischaemic heart disease in 1985 really tells us anything about the spatial distribution of those who suffer from the disease but do not necessarily die from it. It would seem to be a weak base upon which to construct a geographical investigation of illness behaviour and for any spatial analysis of the geography of risk. While medical geography has made a substantial contribution to the study of the spatial patterns of mortality, medical geographers have still been rather parochial in their approach to studying mortality, and the main thrust of their work has been directed towards the portrayal and analysis of the spatial variations of cause-specific mortality and 'the search for environmental and socio-economic conditions which may be causally related to these variations' (Howe and Phillips, 1983, 33). To a large extent they have avoided any detailed consideration of differential mortality, particularly social class and occupational variations, as well as ethnic and racial differences. The same is true of studies of the mortality experience of particular social and demographic groups, such as teenage males, aged pensioners, etc., as well as studies of areally-based social communities such as coal-mining settlements. Even the study of infant and perinatal mortality remains an undernourished field of geographic endeavour.

THE SOCIAL INEQUALITY OF DEATH

Inequality in the face of death remains one of the most distinctive and persisting features of contemporary mortality patterns throughout the world. Inequality in life and death are undoubtedly related, and in many cases reflect basic socio-economic divisions within society. A central preoccupation within geography has been the identification of the spatial dimensions of this inequality, even though few geographers have inquired into the underlying causes of such patterns. Indeed, it remains something of an irony, as Jones points out, that many geographers fail to appreciate fully that mortality patterns are perhaps as well understood through the spatial analysis of social groups as through basic environmental influences (Jones, 1981, 52). This section attempts to trace some of the more important social differentials in contemporary mortality patterns.

Rich Man, Poor Man... Social Class Differentials

The existence of significant variations in mortality levels among various social, economic and ethnic groups has been repeatedly demonstrated both in developing and developed countries. In broad terms, mortality would seem to vary inversely with social class or socio-economic status. Such a circumstance would seem perfectly understandable, given that a high social status is normally associated with better living and working conditions, better nutrition, maternal status and 'mothering', and a greater knowledge of and sensitivity to health-care facilities and preventive medicine (factors which affect individuals), as well as better sanitation, public health and avoidance of environmental hazards (which affect areas). Poor health status (often leading to premature death) is consequently a determinant as well as an outcome of socio-economic conditions. The differing life styles of social groups are also matched by important differences in health behaviour. Most of the evidence on social class differences in mortality comes from studies of developed countries, particularly Britain and the United States of America. In Britain such variations are well documented (see Blaxter, 1976; Brotherston, 1976; OPCS, 1978; Office of Health Economics, 1979) and strongly suggest that social class differences are not restricted to mortality but characterise the whole 'gamut of physical and social attributes upon which estimates are made of normal health development' (Brotherston, 1976, 74). There are, for example, persisting social class differences in 'mothering', birthweights, heights and weights of schoolchildren, eyesight, dental health, nutrition, smoking behaviour, chronic and acute sickness patterns, personal hygiene and comprehension and utilisation of health-care facilities. At the level of mortality such variations can be illustrated via standardised mortality rates by social class groups in England and Wales (Table 4.1). This material shows a strong social class gradient from Class I to V and in addition suggests that, not only has this pattern persisted over time, but also that there has occurred a widening of class differences, rather than a narrowing, over the last 50 years, in particular, those in the unskilled group clearly stand out as suffering considerable disadvantage in terms of health and premature death. Figure 4.1 adapted from the OPCS Report of Occupational Mortality, further highlights the mortality gradient between the various social groups in Britain, both for working-age adults and for children aged from one to fourteen years. Most major causes of death, such as circulatory diseases, neoplasms, and respiratory disorders, also show this social class gradient, particularly when males only are considered. The magnitude of such variations remains striking, and questions the broad assumption that all groups of society have participated equally in the overall decline of mortality that has characterised developed societies over the last 80 or so years. In addition, it argues strongly for a geography that appreciates the spatial distribution of social classes as a prime determinant of the spatial patterns of mortality.

Some of the class differences in mortality are undoubtedly due to the pursuit of occupations which may, by their very nature, be more hazardous in terms of location and work routines and exposure to various environmental threats, including the greater likelihood of accidents and less job security and job satisfaction, as well as restricted access to health care. Figure 4.2 shows mortality rates for Australian males in a series of broad occupational categories. Males employed in professional, executive, administrative and managerial occupations had a mortality experience in 1969-73 substantially below the national average (100), as did males engaged in agricultural activities. Both these groups enjoyed a level of mortality of about half that of the most disadvantaged occupational group (miners, quarrymen and related workers). While the connection between ill health, mortality and occupation is often strong and direct, it is largely overshadowed by other factors related to occupation, such as income, life-style and place of residence, and embraces such things as possession of household amenities, overcrowding, personal health behaviour, diet, smoking, alcohol consumption, and knowledge and use of health-care facilities. Surveys of social class variations in health status and care consistently highlight how social class differences permeate all levels of health and behaviour. Evidence for chronic illness and cigarette smoking clearly reveals a social class gradient. It has also been repeatedly demonstrated that people in the lower social classes are much less likely to respond to health education programs and to attend antenatal, child health and dental clinics, and are less likely to seek medical treatment or avail their children of a variety of preventive measures such as immunisation against the more common infections like diphtheria, polio and measles. Inequal-

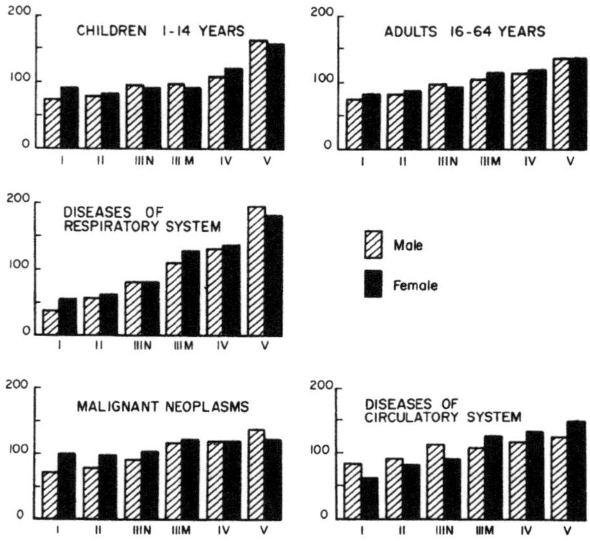

Fig. 4.1 Standardised mortality rates by social class for age groups and selected cause of death, England and Wales, 1970-72. Source: OPCS Occupational Mortality, 1970-72 (1978).

ities of use are compounded by inequalities of access, and a number of studies have vividly revealed how spatial irregularities in the health-care net can lead to a spatial maldistribution in the provision of services which may seriously disadvantage poor people in lower-class neighbourhoods (Shannon and Dever, 1974; Stimson, 1982).

Important social class variations have also been documented in studies of infant and perinatal mortality, and again there is considerable evidence to suggest that, despite a substantial fall in infant mortality rates in developed countries over the last 40 years, the difference between the social classes has remained as persistent as ever. Table 4.2 illustrates the dramatic social class gradient in infant mortality rates that prevails in England and Wales. The wives of men in low-status occupations were about twice as likely to lose their babies in their first year of life as were those in Class I. Similar disparities prevail with respect to perinatal, neonatal and postneonatal mortality. It would also appear that something of a social class gradient exists in birthweight data as well, particularly for the lightest group of babies (under 1500 grams) who suffer two-thirds of the perinatal mortality in the

Table 4.1: Standardised Mortality Ratios by Social Class, males aged 15-64 years England and Wales 1921-72

	Social Class Groups*							
	I	II	IIIN	III	IIIM	IV	V	Ratio I:V
1921-23+	82	94		95		101	125	1.52
1930-33+	90	94		97		102	111	1.23
1949-53+	86	92		101		104	118	1.37
1959-63	76	81		100		103	143	1.88
1970-72	77	81	99	104	106	114	137	1.78

+ 20-64 years

* I Professional occupations (e.g. lawyers, physicians)

II Managerial and lower professional groups (e.g. teachers, nurses, managers, farmers)

III Skilled occupations (e.g. see IIIN and IIIM)

IIIN Non-manual skilled occupations (e.g. clerical workers, shop assistants)

IIIM Manual skilled occupations (e.g. engineering craftsmen, miners)

IV Semi-skilled occupations (e.g. agricultural workers, postman)

V Unskilled occupations (e.g. porters, labourers)

Source: Occupational Mortality 1970-72, 1978; <u>Social Trends</u> 6, 1975:26.

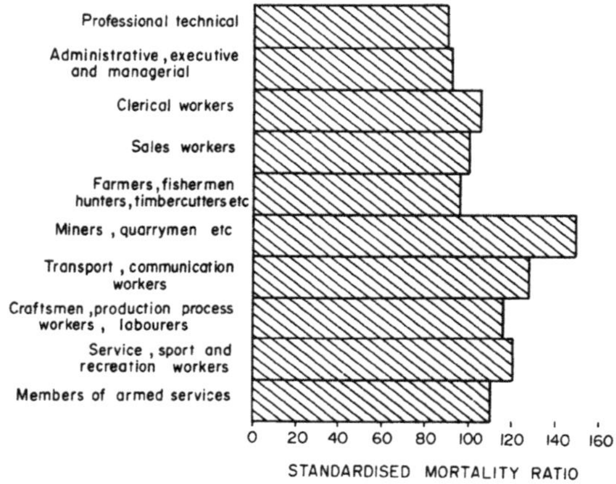

Fig. 4.2 Standardised mortality ratios, Australian males, occupational groups. Source: Population Report, 5, 1981, Australian Council on Population and Ethnic Affairs.

United Kingdom (Office of Health Economics, 1979, 4), even though the pattern is less clear than that for infant mortality. In all cases, the rates for Classes IV and V were significantly higher than those of Classes I and II. It would appear that these broad differences in mortality experience also have a significant spatial expression, as can be seen in Figure 4.3 , which explores the spatial pattern of average infant and perinatal mortality rates for 1979-81 in England and Wales as well as the relationship between the infant and perinatal mortality experience of the two lowest social groups compared to the two highest.

In developing countries, although it is extremely difficult to obtain reliable comparative data, it is widely recognised that social class differentials in mortality, particularly infant mortality rates, are even greater than those recorded for developed nations. A number of studies have indicated a strong association to exist between mothers' education and the level of infant and child mortality, as well as a broad socio-economic gradient in mortality levels (see Brass, 1980; Edmonston and Andes, 1983; Haines and Avery, 1982; Hobcraft et al., 1984). Level of income, for which very little data are available, undoubtedly exercises an important effect on mortality, especially infant and child mortality. Apart from having an obvious effect on housing, diet and sanitation, it also opens up a wide range of health and medical care options, ranging from simple, everyday aspects of health and hygiene, such as personal and household cleanliness and

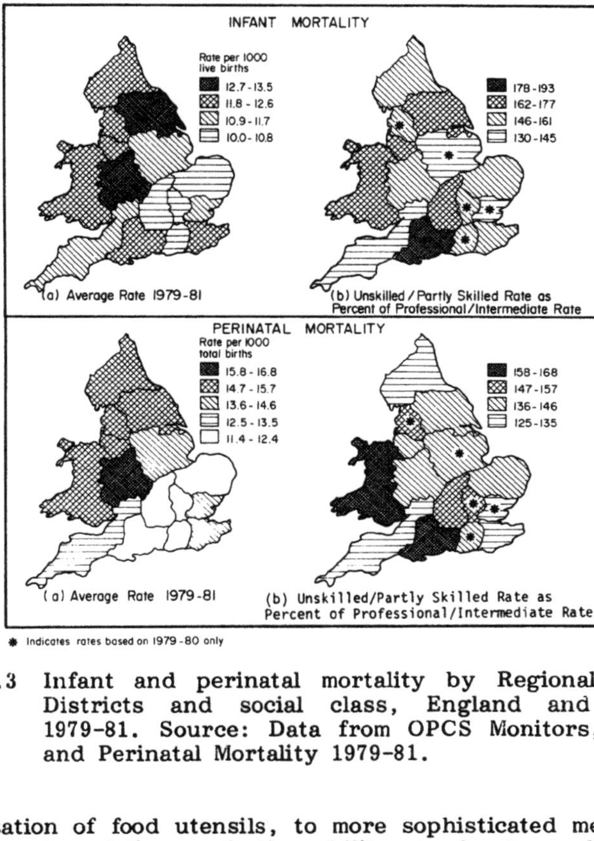

Fig. 4.3 Infant and perinatal mortality by Regional Health Districts and social class, England and Wales, 1979-81. Source: Data from OPCS Monitors, Infant and Perinatal Mortality 1979-81.

sterilisation of food utensils, to more sophisticated measures, such as inoculation and the ability to locate and access modern health care. Hobcraft et al., in documenting some of the socio-economic differentials in child mortality over a wide range of countries, stress the importance of both mothers' and fathers' educational level and fathers' occupation and mothers' work status (all surrogates of household income levels). They also demonstrate the substantial spatial variations that exist in infant and child mortality across the developing world and, in particular, the inequalities that exist within particular countries. Infant and child mortality are influenced by the unequal availability of health and medical facilities and, in particular, the regional disparities that exist between urban and rural areas. Hobcraft et al.'s data dramatically illustrate the wide differences that exist between the health of children belonging to urban 'elite' populations compared with those of the traditional rural

Table 4.2: Infant Death Rates by Social Class England and Wales Average 1979-81. (Average annual rates per 1000 live births)

Social Class Group+	Perinatal*	Neonatal	Postneonatal	Infant Mortality
I	9.4	5.6	3.2	8.8
II	10.9	6.2	2.9	9.0
III	12.7	6.9	3.6	10.5
IV	14.9	8.5	4.9	13.5
V	17.1	9.7	7.1	16.8

* per 1000 total births.
\+ See footnote to Table 4.1 for social class definitions.

Source: Calculated from OPCS Monitors Infant and Perinatal Mortality, 1979, 1980, 1981.

population. On the three measures examined (neonatal, postneonatal and child mortality), the authors found the difference between these two broad socio-spatial groupings to vary from as much as 20 times for child mortality, eight times for neonatal, and six times for postneonatal mortality (Hobcraft et al., 1984, 222-223).

Outsiders - Racial and Ethnic Variations

Some of the most significant differences in mortality patterns in developed countries today are those between racial and ethnic minorities and the majority population. Data on such aspects are often limited, but those which are available indicate that mortality levels, particularly childhood mortality, may be of the same order as found in many countries in the transitional stage of development. Aboriginals in Australia, Negroes and Hispanics in the USA, Asians and West Indians in Britain, and Maoris in New Zealand all suffer from a disadvantaged health status often leading to premature death, despite dramatic improvements in general mortality rates in the last 25 years. In Australia, for example, Aboriginal infant mortality rates remain at least three times greater than the white Australian population, and for parts of the country the difference may be as great as five times. In 1980 the Aboriginal infant mortality rate was estimated to 32.7 compared with a non-Aboriginal rate of only 10.2 (Thomson, 1982, 3). Outside the major cities Aboriginals seem particularly disadvantaged. A 1980-81 study of perinatal outcome among rural Aboriginals in NSW revealed some of the dimensions of this health inequality. Males' life expectancy at birth was estimated to be only 48-49, with that of females being only six to seven years better. The stillborn and neonatal mortality rates were double those of the total NSW population, and the Aboriginal hospitalisation rate two and one half times as high as the non-Aboriginal rate. In addition, the percentage of low-birthweight babies was double the non-Aboriginal rate and, in approximately 30 percent of all confinements, the mother was aged under 20 years, compared with about eight percent of all births in the State. Fourteen percent of Aboriginal mothers were of high parity (four or more children), compared with only four per cent of non-Aboriginal mothers (Julienne et al., 1983). Clearly, Aboriginal health patterns are distinctively different from the non-Aboriginal sector of Australia's population. The factors responsible for such inequality are complex, but Aboriginals are caught up in a vicious cycle of poor health, undernutrition, low educational levels, poverty, unemployment, and inadequate and overcrowded housing with often poor sanitation. In epidemologic terms, what differentiates Aboriginals from non-Aboriginals today is their high level of infectious disease (particularly enteritis and diarrhoea), a high rate of alcohol-related diseases (especially cirrhosis of the liver and pancreatitis), high infant mortality (particularly perinatal and cot deaths), and a high level of pneumonia.

In the USA, racial differences in mortality also persist and probably will remain a problem as long as broad socio-economic differences remain between the black and white populations. Despite a substantial narrowing of the gap between black and white life expectancy in the last 40 years, black Americans today can, on the average, probably expect to live about six years less than white Americans. In addition, black American males have the lowest life expectancy at birth of all the major minority groups in the USA. With respect to infant mortality, the gap between black and white babies has not narrowed at all over the last 50 years. Black babies are almost twice as likely to die in their first year of life compared with white babies, and they are also more likely to weigh less at birth. Black mothers are less likely to seek prenatal care, and are four times as likely to die giving birth compared with white mothers. It would seem that mortality differences are substantially smaller between black and white children from similar socio-economic backgrounds. Most of the differences between black and white children can be accounted for by the higher proportion of black infants born to disadvantaged mothers. Once born, and having successfully survived the first year of life, black children are less likely to be the recipients of health care. In 1979, for example, only 39 per cent of black 1-4 year olds were immunised against polio compared with 64 per cent of white children, and the same situation applied to measles and rubella immunisation (National Center for Health Statistics, 1983, Table 5). More recent figures indicate that almost 31 per cent of all black families (about two million families) existed below the poverty level compared with only nine per cent of white families (Reid, 1982, Table 13). Substantial variations also exist in the causes of death between black and white Americans. Figure 4.4 indicates the trend in mortality rates for eight major causes between 1950 and 1977. Only in the case of deaths by suicide did the white rate exceed that of the black population. In some cases, such as homicide, the black death rate was five times that of the white population.

Evidence from Britain, New Zealand and France indicates that much the same disadvantaged health and mortality status characterises many of the major minority racial and ethnic groups. Gillies et al. (1984) have identified some of the dimensions of health inequality in a study of infant mortality among Asians living in Bradford. This study shows Asians to have had a consistently higher level of stillbirths and of perinatal and infant mortality rates in the period 1975-81 compared with non-Asians. The authors argue that this situation reflects the operation of several inter-related factors, but in the main reflects social class. In Bradford the

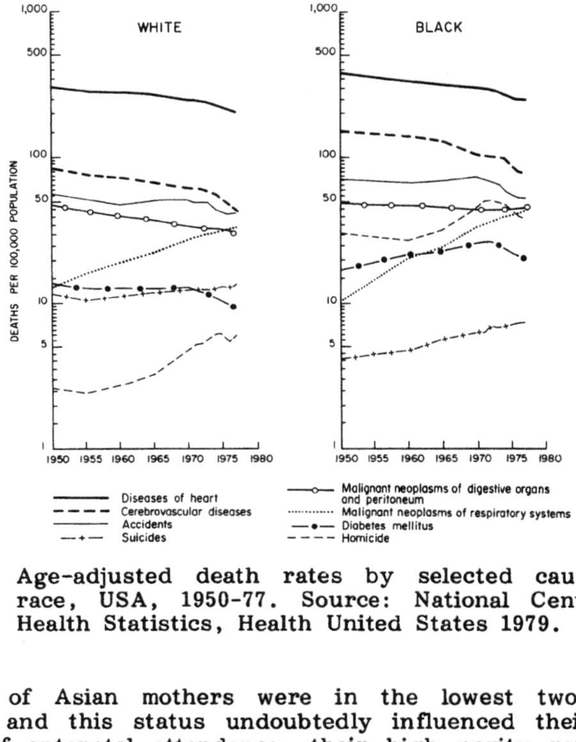

Fig. 4.4 Age-adjusted death rates by selected causes by race, USA, 1950-77. Source: National Center for Health Statistics, Health United States 1979.

majority of Asian mothers were in the lowest two social classes, and this status undoubtedly influenced their poor record of antenatal attendance, their high parity rate, the practice of bearing children until late in life, and the high proportion of low-birthweight babies.

Finally, Hennan and McCracken, in the course of discussing the spatial distribution of mortality in New Zealand in the late 1960s, drew attention not only to the very high differential between Maori and European infant mortality but also to the wide spatial variations that existed with peaks of high Maori mortality in Taranaki, the East Coast, and Thames-Coromandel. Their work adds support to the view of a distinctive spatio-temporal diffusion of mortality control spreading initially from areas where Maori mortality had always been low (largely the major urbanised districts with large Maori populations) outwards to engulf, at a much later date, the more isolated higher mortality areas of traditional Maori settlement (Heenan and McCracken, 1972).

When it comes to assessing the factors responsible for racial and ethnic variations in mortality, it is difficult to go beyond Kitagawa and Hauser's comment that '...race differentials, in general, are consistent with the inverse relationship between mortality and socio-economic status.' (Kitagawa and Hauser, 1973, 102).

Major Causes of Death

For much of the world, the recording of cause of death remains fragmentary and largely unsatisfactory. Fox (1972) has described some of the problems relating to the registration of deaths in the Federal District of Mexico, and questions the way deaths are allocated to their habitual place of residence. Only in the developed countries is a consistent source of cause of death material readily available, and even here there are reasons to call into question the quality and accuracy of many returns. Cause of death registration is often subjective, and is influenced by advances in diagnostic methods as well as fashions in reporting and variations in reporting procedures. One basic problem concerns the differing styles of death certification between developed and developing countries. Some years ago a World Health Organisation study of urban mortality drew attention to these variations, pointing out that 39 per cent of cause of death statements on death certificates in Mexico City may be incorrect, compared

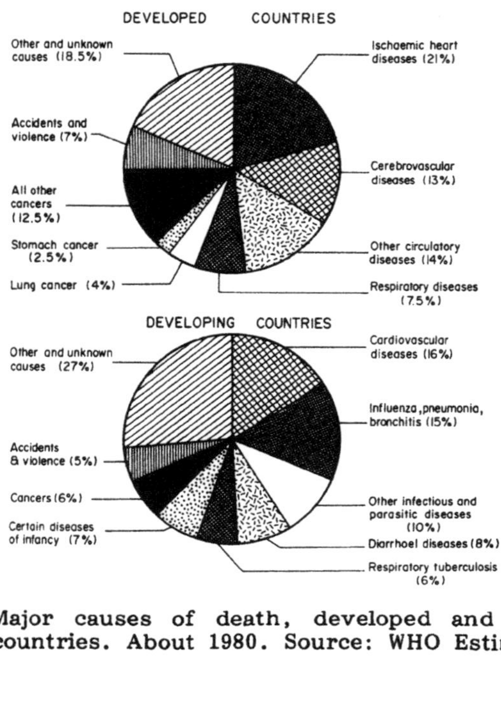

Fig. 4.5 Major causes of death, developed and developing countries. About 1980. Source: WHO Estimates.

with 22 per cent in Bristol and 26 per cent in San Francisco (Puffer and Griffith, 1967). In a more recent study in Atlanta, Engel et al. (1980) found evidence of significant under-reporting of some causes of death (e.g. malignant neoplasms) and over-reporting of others (e.g. vascular disease), as well as many inaccuracies in entries made on the death certificates. Despite such shortcomings, the available material principally from the WHO on the major causes of death, for developed and developing countries, is revealing. The WHO estimates for around 1980, which are given in Figure 4.5 indicate marked differences with respect to mortality structure. The developed world, characterised by an older and ageing population, reveals a Western pattern of degenerative diseases, particularly heart disease, stroke, and other circulatory diseases, which together accounted for 48 per cent of all deaths, with cancers adding another 19 per cent. By comparison, because of their more youthful population structure and because mortality continues to be dominated by child infant deaths, the overall distribution of causes of death in developing countries is dominated by diseases prevalent among the young, particularly parasitic and infective disorders, including gastroenteritis and tuberculosis. Although the differing age structure of developed and developing societies helps explain the distribution of deaths by cause, another important reason is the somewhat belated application of public health and environmental measures in developing countries, so that many infectious and parasitic diseases still flourish. There are also marked differences in underlying morbidity and health patterns between the developed and developing world. Consideration of the pattern of hospital admissions in Victoria, Australia and Papua New Guinea illustrate some of these basic differences. Apart from childbirth, pneumonia, malaria, accidents and gastroenteritis are the leading reasons for admission to hospital and health centres in Papua New Guinea. Hospital admissions in the State of Victoria present an altogether different picture, with cancers, mental illness, heart disease, stroke, bronchitis and urinary disorders dominating the morbidity picture.

Geographers have not shown very much interest in the patterns of cause of death per se, but rather have either prefaced their discussion of the spatial patterns of mortality by a brief comment on cause of death or, in some cases, have disaggregated their material into broad size of place categories, such as Heenan's discussion of differential mortality patterns in New Zealand (Heenan, 1975). This study, which examines among other things selected causes of death for New Zealand's 18 urban areas, the two islands, and non-urban areas, suggests that life in a major city carries with it a greater risk of death from particular forms of heart disease, cancer, and suicide (Figure 4.6). Equally striking is the extent to which areas outside the main urban centres are significant for other forms of heart disease, infections, pneumonia and diabetes. The study also suggests a substantial variation in the risk of cancer by place of residence.

THE SPATIAL PATTERNING OF MORTALITY

So far, this chapter has considered only some of the major social differentials in mortality. To a large extent these reflect variations in life style, nutrition, work patterns, attitudes to health, and personal behaviour. Geographers have, by contrast, paid more attention to mortality differentials between different places. The geographic approach in studying mortality has largely been to map static patterns of mortality, to describe the resulting distributions and, in some cases, to test their significance. Where causative analysis has been attempted it has, for the most part, ranged from essentially subjective visual comparisons for two or more spatial patterns to more sophisticated statistical methodologies extending from simple correlation, multiple correlation and

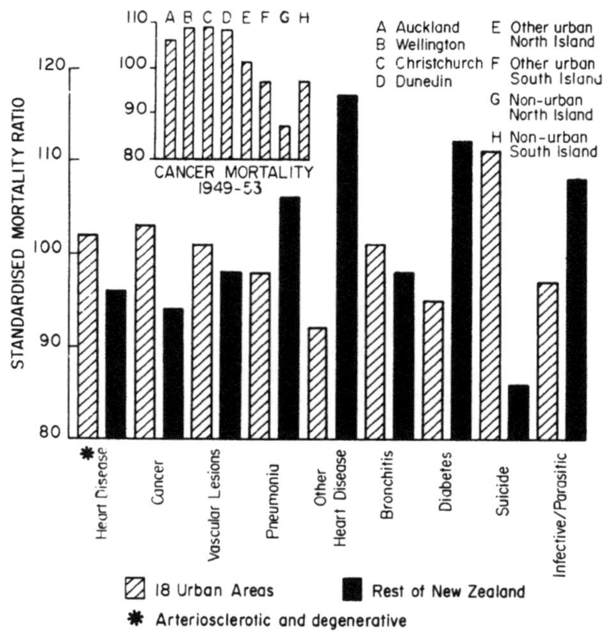

Fig. 4.6 Standardised mortality ratios by urban and non-urban areas. New Zealand, 1964-67. Source: Data from Heenan (1975) Tables 1 and 2.

regression analysis to those which undertake an initial screening and distillation of independent variables before regression methods are used to search out underlying causal determinants (Wilson, 1978, 150).

The utility of such an approach rests on the belief that both death and disease display distinctive spatial patterns, and that the identification of these may lead to hypotheses as to cause. More specifically, the identification of areas of high mortality, or of areas where mortality from a specific cause is especially prevalent, can facilitate health-care planning, maximise the effect of particular health programs, and avoid inequalities in the provision of health care. Equally important, the identification of the spatial contours of mortality may a) illuminate broad social and spatial variations in the human condition, and b) provide clues as to cause and thus to prevention. Many geographers, therefore, see the identification of the spatial pattern of mortality as representing a first step in serious causal inquiry. At a slightly lower level of expectation, the coincidence of particular environmental and socio-economic factors with particular spatial patterns of mortality may suggest fruitful lines of further epidemiologic investigation. Finally, the delineation of geographical areas affected by high and statistically significant levels of mortality may pinpoint particular components of the physical and social environment hazardous to health that are not normally apparent. If anything, most of the geographic research investigating mortality has been carried out within the tradition of the pattern-orientation approach (Woods, 1979). This work has been primarily cross-sectional in approach, broadly descriptive, and concerned mainly to identify spatial patterns as an endpoint in themselves, rather than with any vigorous search for causes pursued at micro level. In addition, nearly all the geographic studies of mortality have proceeded, by employing residentially-based data, to produce maps of mortality on the basis of where people resided at the time of death. Although it is true that geographers remain prisoners of their data sources and, in particular, of the spatial net for which data are collected, it would seem that a strong case can be made for mortality maps and analyses based on workplace locations, residence and job history, and nature of employment. Almost all geographic studies have relied on the use of standardised mortality rates often calculated for males and females separately. The value of this summary measure has recently been questioned by McCracken who, in a short but stimulating paper, argues the case for disaggregating mortality into age-specific categories. McCracken points out how standardised mortality rates may obscure important age-specific variations in mortality (McCracken, 1981). Choropleth maps are the most widely used descriptive method and have the advantage of geographical precision and the disadvantage of placing more visual weight on the large, more sparsely or unevenly populated spatial units while failing to give emphasis to smaller areas such as towns and cities. In an effort to surmount this problem several geographers, such as Howe(1970) and Heenan (1976), have employed more stylised demographic formats in which the population size of the areal unit is made proportional to the total population at risk.

International Contrasts

There is a long tradition in geography concerned with observing, mapping and describing worldwide patterns of mortality distribution, such as crude death rates, life expectancy, and infant mortality. Such work fits within the pattern-approach of population geography, and is primarily preoccupied with considerations of Where? rather than with Why there? (Jones, 1981, 6). Although many geographers have examined inter-national patterns of crude death rates, such a measure conceals wide differences between regions and countries, and has no real public-health meaning. Expectation of life at birth provides a better insight into the current health status of population, being largely independent of age structure, even though it is not age-standardised. Figure 4.7 shows the spatial distribution of life expectancy at birth around 1980. An examination of this map reveals a variety of mortality patterns. Those countries where life expectancy is 50 or below are distinctively equatorial or tropical in distribution. Almost all of Africa south of the Sahara and north of the Republic of South Africa falls within this disadvantaged group, as does a group of south and southeast Asian countries extending from Afghanistan through India, Nepal, Bangladesh, Kampuchea and Laos to Indonesia. By contrast, western and northern Europe, North America, Japan and Australasia all enjoy a life expectancy at birth in excess of 70 years and, in the case of Australia, Japan, Sweden, Norway and Switzerland, in excess of 75 years. Most of Latin America, with the exception of the Argentine, Uruguay and Haiti, and the Islamic world had an intermediate expectation of life of between 55 and 64 years, while the USSR and China had life expectancy levels of between 65 and 69 years.

The most important difference between the mortality experience of developed and developing countries quite clearly occurs in the first year of life (Figure 4.8). North and western Europe, North America, Japan and Australasia vividly stand out as a region of very low infant mortality. At the other end of the spectrum, in much of Africa, the Middle East and South Asia infant mortality rates are at their highest, often in excess of 140 deaths per 1000 live births. Between these two extremes lies a handful of east and south European States, together with Cuba and Chile, with rates between 12 and 25. Finally, much of central and South America, China,

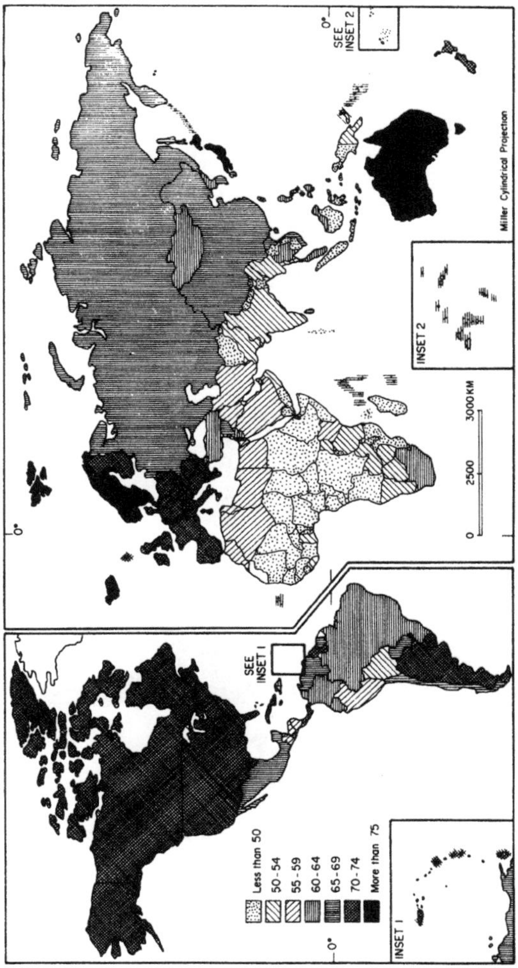

Fig. 4.7 Life expectancy at birth about 1980. Source: Data from Population Reference Bureau, World Population Data Sheet, 1984.

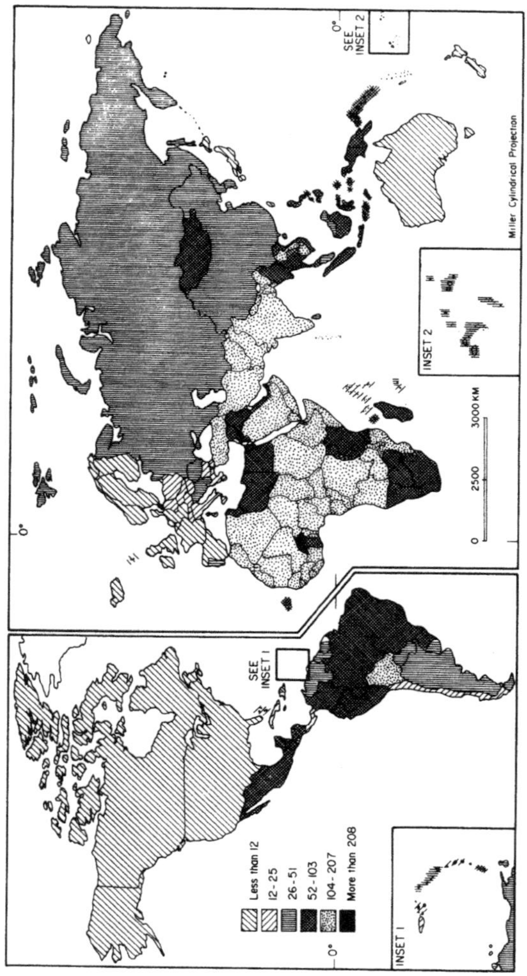

Fig. 4.8 Infant mortality rates about 1980. Source: Data from Population Reference Bureau, World Population Data Sheet, 1984.

the USSR, Yugoslavia and Romania have intermediately high rates of between 26 and 103 deaths per 1000 live births.

Regional Contrasts

A number of geographers have addressed the problem of regional inequalities in mortality levels, both from the point of view of demonstrating how national figures can disguise wide regional variations in health and mortality, and in an effort to discover how such differentials change over time. Carvalho, for example, provides estimates of life expectancy rates in 10 regions of Brazil, and documents a significant decrease in regional variation between 1930 and 1970 (Carvalho, 1974). Field, in his broad overview of Canadian regional mortality patterns over the last half-century, also finds evidence of a decline in the magnitude of mortality differentials (Field, 1980). In perhaps the widest-ranging study, Van Poppel discusses the pattern of male and female life expectancy for almost 260 regions of north and western Europe in 1969-77. His work demonstrates a clear link between areas of relatively low male life expectancy and the location of urbanisation, mining, heavy industry, and dockyards. By contrast, agricultural areas and the more affluent regions of urban/suburban settlement and economic development, such as in southeast England, possess a much more favourable male life expectancy (Figure 4.9). In Britain there also appears a well-marked gradient from the older industrial areas of the north and northwest to the more advantaged south and east (Van Poppel, 1981).

Although substantial regional differentials have been found to exist in mortality, and although there is some evidence that, over time, such differentials lessen in magnitude, it would appear the the trend is not as marked for all age and social groups. Knox, for example, in examining regional patterns of infant mortality in Britain from 1949 to 1972, finds evidence of considerable stability in patterns of social and spatial inequality. Commencing with the assumption that social class gradients in infant mortality have persisted and perhaps widened in this period, Knox was interested in discovering whether or not such inequalities had spatial expression, and whether spatial gradients in infant mortality have parallelled social class gradients over the last 20 years. In fact, he discovered that the pattern of spatial inequality changed very little between 1949 and 1972. Overall, the spatial distribution of infant mortality tended to reflect long-established variations in the distribution of social classes and levels of unemployment, and variations in the quality of life and environmental conditions, all of which may have had an additional effect on variations in the quality and utilisation of infant health care. Knox concluded that patterns of infant health both reflect and bolster the distribution of higher status

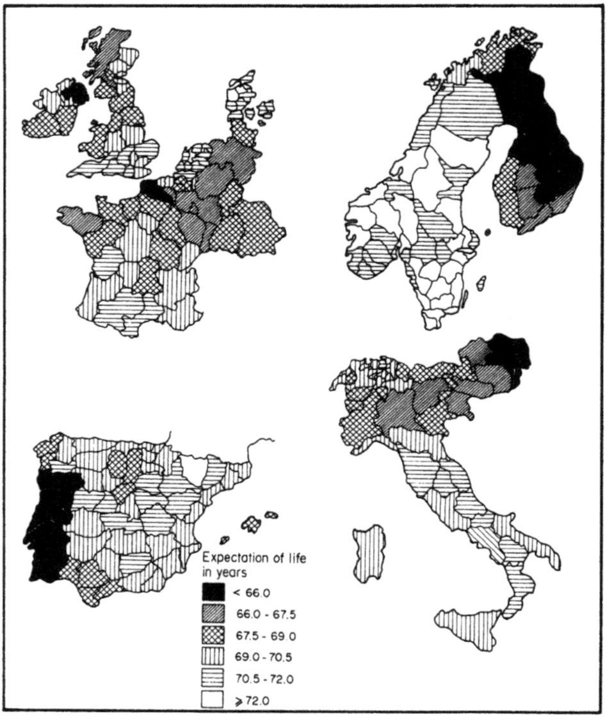

Fig. 4.9 Male life expectancy at birth, regions of Western and Northern Europe 1969-77. Source: Van Poppel (1981).

groups in society at the expense of the more disadvantaged groups, and that this spatial pattern was both long established and persistent (Knox, 1981).

In a study of regional and community variations in infant mortality in Peru, Edmonston and Andes found evidence of a substantial rural-urban differential in mortality, and suggest that mortality is inversely related to community size, with the highest rates in mountain Sierra communities and the lowest in metropolitan Lima. Although there was a strong association between socio-economic status, mothers' educational level and infant mortality, altitude also displayed a persistently positive association with community mortality, even after controlling for socio-economic status and the distribution of health facilities (Edmonston and Andes, 1983). Turning now to regional analyses of cause-specific mortality patterns, Howe (1970,

1976) has provided a series of detailed cause-specific mortality maps of the United Kingdom, Murray (1967) maps of mortality patterns in the USA, Heenan (1975) maps of differential mortality in New Zealand, and McGlashan (1977) a series of maps showing mortality rates from selected causes in Australia. Most of these studies have worked with data collected at the broad statistical level of counties, shires, boroughs, towns and cities, and all have used standardised mortality rates to identify areas of high and low mortality from specific causes. Howe, for example, shows how high rates from heart disease are concentrated in the north of the United Kingdom, in south Wales, and in parts of Lancashire, Yorkshire, Teeside, Scotland and Northern Ireland, as well as in several London boroughs. The coincidence of heart disease with densely settled industrialised districts and with lower-class residential areas of the major cities is clearly demonstrated by this work, and by the work of McGlashan (1977) and Burnley (1982) in Australia. In recent years, a number of geographers have focused on the increasing toll of unnecessary death associated with the pace of modern urban living, such as alcohol-related deaths, motor vehicle accidents, and suicide. McGlashan has, for example, examined the spatial variations in alcohol-related deaths in Tasmania, and finds a significant statistical relationship to exist between cirrhosis of the liver deaths, alcoholism, and the masculinity and marriage rate (McGlashan, 1980). Burnley and McGlashan have also analysed the spatial pattern of suicide deaths in Australia, using a variety of mapping conventions (Burnley and McGlashan, 1980), and Brodsky and Hakkert have examined the spatial distribution of motor vehicle deaths in the rural areas of Texas in terms of access to emergency medical services (Brodsky and Hakkert, 1983). Selya also examines motor vehicle deaths, but in this case within the island of Taiwan and within the framework of the epidemiologic transition model. He raises the interesting question as to whether there is evidence of significant changes in risk-taking behaviour associated with societies undergoing rapid social change, and links this to the increase in fatal road accidents (Selya, 1980). A number of Japanese scholars have studied the regional distribution of deaths from cerebrovascular disease (Takahashi, 1981; Tamashiro et al., 1981) and lung cancer (Minowa et al., 1981). Most of this work has been directed towards the aim of providing hypotheses for the investigation of cause, such as Takahashi's discussion of the geographical patterns of cerebro-vascular deaths in Japan, which revealed a strong northeast-southwest gradient, with the highest mortality rates in the northeast prefecture of the main island (Tohoku region), and the lowest in the southwest prefectures surrounding the inland sea, as well as on the island of Hokkaido. Within the Tohoku region, Takahashi also found considerable variation, particularly between the coastal fishing communities and the inland farming communities. This led the author to probe variations in the pattern of diet and environmental factors such as salt intake (very high in the Tohoku region), ambient temperatures, and the biochemical environment of the islands. More recently, Mohan and Rhind, in a monograph perhaps more important for the method it advances than for the results of its analysis, have examined small-scale variations in mortality experience within Durham County in England. Using the signed Chi-square statistic, which takes into account absolute and relative deviations from the norm as well as permitting a quasiprobabalistic approach to mortality mapping, they examine patterns of cause-specific mortality at varying levels of spatial resolution (Mohan and Rhind, 1983).

Intra-City Contrasts

The application of geographical methods to mortality patterns has also proved fruitful at the metropolitan level, even though there are still relatively few studies available, particularly of cities in the developing world. Most of the studies available are static single-city studies, and there have been few attempts either to trace the changing spatial patterns of mortality at the most detailed ecological level, or to provide comparative analyses of the mortality experience of several cities. One of the few examples of the latter approach is Howe's discussion of the distribution of ischaemic heart disease and cancer deaths in London and Glasgow (Howe, 1972, 1982). In these brief papers Howe illustrates the considerable spatial variation in mortality patterns that can exist between and within large British cities sharing the same socio-economic and health-care system, as well as the same broad avenues of access to doctors and hospitals. Whereas the pattern of mortality from heart disease and cancer showed a declining incidence as one moved outward from the core in London, in Glasgow much more of a sectoral pattern was evident. Howe indicates the population groups most at risk from premature death from cancer and/or heart disease to be those living in the industrial lower-status suburbs.

The majority of studies investigating intra-urban patterns of mortality remain broadly descriptive in design, seeking to identify, map, describe and explain urban inequalities in mortality rates and relate these variations to the spatial distribution of a variety of environmental and socio-economic variables. Most studies are couched at the broad aggregative level, relying on data collected for broad administrative areas such as boroughs, municipalities, and electoral districts. The most common method has been to map standardised mortality rates for particular causes (particularly heart disease, cancers, cerebrovascular disease and suicide), describe the pattern, in some cases test it for significance, and then to loosely hypothesise possible risk factors either intuitively or from comparison with other distributional patterns. An additional step has often been to assemble an array of environmental and socio-economic variables in a correlation matrix and take the resulting positive and negative associations with mortality as indicating causality. Burnley has, for example, analysed mortality patterns in Sydney and Melbourne (Burnley, 1977, 1980, 1982), Gibson and Johansen (1979) in Sydney, Howe (1970, 1972, 1979) in London and Glasgow, Moens (1984) in Brussels, Pyle (1971) and Pyle and Rees (1971) in Exeter, while Fox (1972) provides one of the very few studies of intra-urban mortality patterns in a major city of the developing world. Most of these studies have consistently reported high rates of mortality to exist in inner-city areas, in blue-collar suburbs, and in and around industrial areas. This seems to apply to a wide variety of causes of death, but particularly to total mortality, infant mortality, ischaemic heart disease, some cancers, peptic ulcer, cirrhosis of the liver, deaths from infectious disease and suicide. Exceptions to the above appear to be motor vehicle accidents and stomach cancer. A number of studies have also considered the problem of the spatial patterns of infant mortality within the city, such as Wilson (1972, 1979), Burnley (1977), Wood (1982), and Picheral (1976). Wilson, in his study of infant deaths in Wollongong, Australia, suggest that the uneven spatial distribution of infant mortality within the city could be accounted for by a cluster of variables representing family status, maternal characteristics, ethnicity, household characteristics, poverty, unemployment, and environmental factors. In particular, he postulates the existence of a direct causal link between non-standard families and infant deaths operating through the medium of such things as diminished mothering, poor antenatal care, poorer maternal conditions (for example, low age at birth, ethnic status, illegitimate birth) or, in some cases, deliberate maltreatment (Wilson, 1979).

THE IDENTIFICATION OF RISK FACTORS AND CAUSAL GEOGRAPHIC RESEARCH

As mentioned above, many geographers have claimed that the identification of spatial patterns of mortality represent the first stage in a sequence of more comprehensive investigations leading to cause. To a large extent the effort has been to discover those components of the natural and social environment hazardous to health and responsible for premature death. The problem of establishing causality would seem to be a central issue for the geographic study of mortality. Probably the scientific demonstration of causality in geographic research remains an illusory and unobtainable goal. The basic problem is that statistical associations or correlations may be suggestive, but never conclusive, proof of a cause-effect relationship. In practice this means that the matter of determining causality from correlation analyses is heavily dependent upon the subjective biases of the individual researcher. Given this, many of the claims of medical geography that the spatial approach may unearth causal patterns need careful scrutiny. Much work in geography has endeavoured to establish causes by variable association or associative occurrence of a number of environmental and socio-economic factors, and in many cases the results have been bio-socially improbable and pay little attention to confounding or intervening factors. In addition, the inference that residential location plays a role in the cause - effect sequence governing human mortality needs questioning. It is quite possible, for example, as Smith remarks, 'that mortality patterns may be more an outcome of personal or group circumstance that of local or environmental quality' (Smith, 1982, 3). The problem of what King has called latency and mobility also has not been fully appreciated by geographers (King, 1979). Much research on mortality does not take into account the time-lag between exposure to a particular disease agent and death. Degenerative diseases may not lead to death until several decades after the initial onset, and in the interim the environmental and/or socio-economic factors responsible for the symptoms of the disease may have vanished. This has wide-ranging implications for a geography of mortality based on the assumption that place of residence at death is in some way a factor leading or contributing to death. It would seem that, rather than pursue residence-based studies of mortality, it would be much more fruitful to employ longitudinal and cohort studies, or perhaps attempt to reconstruct, retrospectively, individuals' residential and work history and their history of exposure to particular environmental hazards. In addition, the range of variables which geographers have utilised in their analyses has been somewhat restrictive. Few attempts have been made to consider the spatial patterning of biochemical and/or physiologic variables or how their distribution may influence the geographical pattern of mortality. One such attempt has been by Gibson and Johansen (1979) who, in addition to providing a number of maps of mortality patterns from particular causes, also include a series of provocative maps illustrating a wide range of bio-social data (see also Meade, 1983). Figure 4.10 presents just three of these maps showing the spatial distribution of diastolic blood pressure, cholesterol levels, and levels of obesity within Sydney.

The Identification of Particular Risk Factors

Most geographers working in the field of human health and mortality have attempted to identify those components of the physical and socio-economic environment thought to be in some way hazardous to human health and possibly linked to premature mortality.

The association between climate and mortality is beginning to attract more attention from geographers, although a vigorous scientific investigation of the relationship has yet to be undertaken. Gentilli (1980) has provided a preliminary but stimulating comment on the influence of climatic factors on Australian health, ranging across a discussion of infection, vector-borne and degenerative diseases. Seasonal rhythms in mortality have also attracted some attention from geographers, with Kevan (1980a, 1980b) discussing seasonal variations in suicide, McGlashan and Grice (1983) the link between daily and monthly minimum temperatures and sudden infant death syndrome, Sato (1981) the seasonality of stomach cancer in Japan, and Paulozzi (1981) seasonal patterns of mortality in Alaska. In urban climates, mortality has been most regularly studied in relation to extremes of temperature. In the USA high mortality rates have been shown to be associated with summer heat-waves in several metropolitan areas. On the average, high ambient temperatures are associated with the deaths of more than 200 Americans every year and produce a distinctive pattern of intra-urban mortality and morbidity. Inner-city areas, together with suburbs of low socio-economic status, black people, Hispanics, the elderly, the physically handicapped and the poor would all seem to be at risk (Schuman, 1972).

The effects of air pollution on human health and mortality have also been long recognised. Currently there is a consensus that the habit of smoking represents a significant risk factor in premature mortality from a variety of diseases, yet very few geographers have considered smoking behaviour in their discussion of cause-specific mortality, with the exception of a recent study by Heenan into the general incidence of smoking in New Zealand (Heenan, 1983). As well as self-imposed air pollution, people's lungs are also exposed to air polluted by the emissions of motor vehicles, industry, and domestic activities. MacDonald, in his study of cancer deaths in Houston, found an association between areas of high air pollution and cancer rates, particularly among the non-white and Hispanic population (MacDonald, 1976). Lead has claimed special attention as a pollutant, mainly because of its toxicity and its fairly regular occurrence in the natural environment. In recent years, considerable attention has been directed towards the level of lead ingestion and the detrimental effects of lead poisoning on young children. Hunter has, for example, examined the summer peak found in lead poisoning in urban-industrial areas of the USA (Hunter, 1978). Both he and Caprio et al. (1975) have discussed the spatial variations in lead absorption in terms of residential proximity to major arterial routeways. The adverse effects of water pollution are also well documented, particularly for

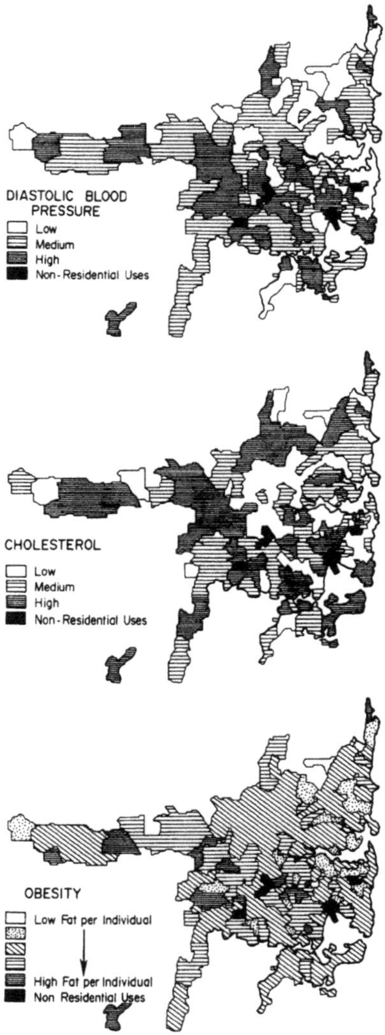

Fig. 4.10 The distribution of diastolic blood pressure, cholesterol and obesity rates, Sydney Metropolitan Area, 1970-75. Source: Simplified from Gibson and Johansen (1979) with permission.

nineteenth-century British and American communities. More recently Beresford (1981), in a retrospective study, has investigated the association between the re-use of drinking water and the level of gastrointestinal and urinary cancer mortality in the London area. Among other things, this study found that the broad relationship was not independent of socio-economic factors which accounted for most of the statistical association (Beresford, 1981).

The physical condition and age of housing, neighbourhood deterioration, and the lack of physical and social amenities have also been examined as a risk factor in human mortality. The role of dampness, inadequate sanitation, dilapidation and overcrowding have long been recognised as having some connection to human health, although a direct cause-effect relationship has been very difficult to prove as a number of related social and economic factors may intervene. The occupants of the most impoverished housing in any community are also likely to be at a disadvantage when it comes to income, nutrition, education, and access to health care. They are also more subject to unemployment and, if they do work, they are more likely to be engaged in unhealthy and hazardous occupations. A number of studies have examined the role of housing in human health patterns. Robinson found, for example, that disadvantaged housing condition was one of two good indicators of heart-disease mortality in north-eastern Ohio (Robinson, 1978).

An area of considerable interest to geography has been the possible association between elements of the geochemical environment and human mortality, even though so far most results have proved inconclusive. The relationships between the physical and chemical components of drinking water and heart disease and cancer have been examined at a variety of levels. Studies from various parts of the world have indicated that soft-water supply is possibly a risk factor in heart disease (Crawford et al., 1968; Howe, 1976; Meade, 1983). Armstrong particularly has pioneered the geographical investigation of the association between trace elements in the soil and cancer mortality, primarily by way of the complex food chain extending from plants and animals through the cultural processes of food preparation (Armstrong, 1967, 1971).

As most studies of mortality patterns by geographers have been carried out within the context of the developed world, with its accent on degenerative and man-made causes of death, it is not surprising that geographers have paid considerable attention to the social factors in premature mortality. Most studies have utilised a broad ecological approach, selecting socio-economic variables from census records usually to represent social class, housing status, social disorganisation, crowding, and ethnic minority status. More recently, geographers have begun to use a much finer level of spatial resolution to identify micro-patterns and to escape some of the problems attendant on the use of large areal units (see, for example, Mohan and Rhind, 1983). The evidence provided by these studies suggests that social class remains one of the strongest determining factors of spatial inequalities in mortality.

Studies of mortality patterns in the developing world have largely placed emphasis on the role of the physical environment, particularly as it relates to a wide variety of infectious and communicable diseases. Mukerjee has, for example, examined the unique ecology of a small Indian region in terms of the environmental determinants of a range of infectious diseases such as malaria and typhoid (Mukerjee, 1980), Dutt et al. the role of the physical environment and land-use practices in an India-wide study of malaria deaths (Dutt et al., 1980), and Kloos and Thompson the role of environmental factors in the incidence and spread of trypanosomes and schistosomiasis (Kloos and Thompson, 1979). The disadvantaged nutritional and health status and the importance of premature death among city slum, squatter and minority dwellers have also attracted comment, with studies by Basta (1977) and Hughes and Hunter (1970). More recently a number of geographers have become interested in the interrelationships between the development process, particularly the modification of the natural environment and the increase in disease hazards and premature mortality. Weil shows how agricultural colonisation in Latin American has led to the altering of the balance between disease agents, vectors and their natural hosts, and greatly increased the transmission rate of many diseases (Weil, 1981). Following up the same broad theme, Meade has shown, in peninsular Malaysia, how voluntary resettlement programs have increased a wide range of disease hazards (Meade, 1978). Water-development projects, such as dams and irrigation systems, have in Africa, Asia and Latin America led to an increase in water-borne and associated diseases, such as schistosomiasis, malaria and filariasis, and created whole new areas of transmission (Hunter et al., 1982). More recently, Sindiga clearly demonstrates how the spread of tsetse and sleeping-sickness deaths in Maasailand, Kenya, was closely related to the colonial administration's policy governing resettlement and game reserves (Sindiga, 1984).

THE FUTURE

There is a growing awareness of the need for geographers to be more concerned with causal processes rather than with straightforward descriptive and correlational analyses. It is also clear that research into the processes which produce spatial differences in mortality will require a higher level of methodological sophistication than has hitherto been the case.

Much more care is required in designing studies that will address the problem of exposure and outcome in a wide variety of environmental and cultural settings, perhaps through the use of retrospective and prospective research strategies of particular populations, groups and individuals. The use of census variables as surrogates can no longer be justified, nor can broad aggregative analyses at the macro spatial level. Specifically, geographers also need to question their preoccupation with residentially-based data and their belief in the importance of locality as a factor influencing premature death. There is also a need to move away from static cross-sectional studies and to provide more comparative and dynamic investigations of mortality. Finally, the neglect of mortality by population as opposed to medical geographers is serious and indefensible. The balance sheet needs to be made more favourable.

REFERENCES

Armstrong, R.W. (1967) 'Milk and Stomach Cancer in Iceland', Acta Agriculturae Scandinavica, 17, 30-21

Armstrong, R.W. (1971) 'Medical Geography and its Geologic Substrate', in H.L. Cannon and H.C. Hopps (eds.), Environmental Geochemistry in Health and Disease, Geological Society of America, Boulder, 211-219

Australian Bureau of Statistics (1984) Public Hospital Morbidity, Victoria 1982, Government Printer, Canberra

Australian Council on Population and Ethnic Affairs (1981) Population Report No. 5, Canberra

Basta, S.S. (1977) 'Nutrition and Health in Low Income Urban Areas of the Third World', Ecology of Food and Nutrition, 6, 113-114

Beresford, S.A.A. (1981) 'The Relationship Between Water Quality and Health in the London Area', International Journal of Epidemiology, 10, 2, 103-115

Blaxter, M. (1976) 'Social Class and Health Inequalities', in C.O. Carter and J. Peel (eds.), Equalities and Inequalities in Health, Academic Press, London, 111-125

Brass, W. (1980) 'Policies for the Reduction of Mortality Differentials', Population Bulletin of ECWA, 19, 3-27

Brodsky, H. and Hakkert, A.S. (1983) 'Highway Fatal Accidents and Accessibility of Emergency Medical Services', Social Science and Medicine, 17(11), 731-740

Brotherston, J. (1976) 'The Galton Lecture: 1975. Inequality: Is it Inevitable?', in C.O. Carter and J. Peel, (eds.), Equalities and Inequalities in Health, Academic Press, London, 73-104

Burnley, I.H. (1977) 'Mortality Variations in an Australian Metropolis: The Case of Sydney', in N.D. McGlashan (ed.), Studies in Australian Mortality, Environmental Studies Occasional Paper 4, University of Tasmania, 29-61

Burnley, I.H. (1980) 'Social Ecology of Premature Mortality in Three Australian Cities', Australian Journal of Social Issues, 15, 4, 306-320

Burnley, I.H. (1982) Population, Society and Environment in Australia, Shillington House, Melbourne

Burnley, I.H. McGlashan, N.D. (1980) 'Variations of Suicide Within Australia', Social Science and Medicine, 14D, 215-224

Caprio, R. J. et al. (1975) 'Residential Location, Ambient Air Lead Pollution and Lead Absorption in Children', The Professional Geographer, 27, 37-42

Carvalho, J.A.M. (1974) 'Regional Trends in Fertility and Mortality in Brazil', Population Studies, 28, 401-421

Clarke, J.I. (1972) Population Geography (2nd Edition), Pergamon, London

Clarke, J.I. (1984) 'Geography and Population: Approaches and Applications, Pergamon, London, 1-10

Crawford, M.D. et al. (1968) 'Mortality and Hardness of Local Water Supplies', The Lancet, 7547, 827-831

Dever, G.E.A. (1972) 'Leukaemia and Housing: An Intra-Urban Analysis', in N.D. McGlashan (ed.), Medical Geography: Techniques and Field Studies, Methuen, London, 233-245

Dutt, A.K. et al. (1980) 'Malaria in India with Particular Reference to Two West-Central States', Social Science and Medicine, 14D, 317-330

Edmonston, B. and Andes, N. (1983) 'Community Variations in Infant and Child Mortality in Peru', Journal of Epidemiology and Community Health, 37, 121-126

Engel, L.W. et al. (1980) 'Accuracy of Death Certification in an Autopsied Population with Specific Attention to Malignant Neoplasms and Vascular Diseases', American Journal of Epidemiology, 111(1), 99-112

Field, N. (1980) 'Temporal Trends and Spatial Patterns of Mortality in Canada', in F. A. Barrett (ed.), Canadian Studies in Medical Geography, Geographical Monographs No. 8, Dept of Geography, York University, Ontario, 31-58

Fox, D.J. (1972) 'Patterns of Morbidity and Mortality in Mexico City', The Geographical Review, LX11(2), 151-185

Gentilli, J. (1980) 'Some Climatic Factors in Australian Health', Social Science and Medicine, 14D, 85-100

Gibson, J.B. and Johansen, A. (1979) The Quick and the Dead: A Biomedical Atlas of Sydney, Reed, Sydney

Gillies, D.R. et al. (1984) 'Analysis of Ethnic Influence on Stillbirths and Infant Mortality in Bradford 1975-81',

Journal of Epidemiology and Community Health, 38, 214-217
Griffiths, M. (1971) 'A Geographical Study of Mortality in an Urban Area', Urban Studies, 8, 111-120
Haines, M.R. and Avery, R.. (1982) 'Differential Infant and Child Mortality in Costa Rica: 1968-1973', Population Studies, 36(1), 31-43
Health Commission of NSW (1979) 'Aboriginal Mortality in Country Areas of NSW 1978, 1979', Working Paper No.2, Sydney
Heenan, L.D.B. (1975) 'Some Spatial Aspects of Differential Mortality in New Zealand', New Zealand Geographer, 31(1), 29-53
Heenan, L.D.B. (1976) 'Spatial Patterns of General and Cause Mortality on the West Coast, New Zealand', New Zealand Geographer, 32(2), 139-159
Heenan, L.D.B. (1983) 'Cigarette Smoking Among New Zealanders: Evidence from the 1976 Census', in N.D. McGlashan and J.R. Blunden (eds.), Geographical Aspects of Health, Academic Press, London, 241-255
Heenan, L.D.B. and McCracken, K.W.J. (1972) 'On the Spatial Distribution of Mortality in New Zealand', New Zealand Medical Journal, 75, 194-200
Hobcraft, J.N. et al. (1984) 'Socio-Economic Factors in Infant and Child Mortality: A Cross-National Comparison', Population Studies, 38, 2, 193-223
Howe, G.M. (1970) National Atlas of Disease Mortality in the United Kingdom, Nelson, London (Second Edition)
Howe, G.M. (1972) 'London and Glasgow: A Comparative Study of Mortality Patterns', in W.P. Adams, and F.M. Helleiner (eds.), International Geography, volume 2, University of Toronto, Montreal, 1214-1217
Howe, G.M. (1976) 'Environmental Factors in Disease', in J. Lenihan, and W.W. Fletcher, (eds.) Health and the Environment, Blackie, London
Howe, G.M. (1979) 'Death in London', The Geographical Magazine, L1(4), 284-289
Howe, G.M. (1982) 'London and Glasgow: A Spatial Analysis of Mortality Experience in Contrasting Metropolitan Centres', Scottish Geographical Magazine, 98(2), 119-127
Howe, G.M. and Phillips, D.R. (1983) 'Medical Geography in the United Kingdom, 1945-1982', in N.D. McGlashan and J.R. Blunden (eds.), Geographical Aspects of Health, Academic Press, London, 33-52
Hughes, C.C. and Hunter, J.M. (1970) 'Disease and Development in Africa', Social Science and Medicine, 3, 443-493
Hunter, J.M. (1978) 'The Summer Disease-Some Field Evidence on Seasonality in Childhood Lead Poisoning', Social Science and Medicine, 12D, 85-94
Hunter, J.M. et.al. (1982) 'Man-Made Lakes and Man-Made Diseases', Social Science and Medicine, 16(11), 1127-1145

Jones, H.R. (1982) A Population Geography, Harper and Row, London

Jones, H.R. (1984) 'Population Geography in Britain', in J.I. Clarke (ed.), Geography and Population: Approaches and Applications, Pergamon, London, 171-178

Julienne, A. et.al. (1983) 'A Comparative Study of Perinatal Outcome Among Aboriginal and Non-Aboriginal Hospital Confinements in Rural NSW 1981', Working Paper No.3, NSW Department of Health, Sydney

Kevan, S.M. (1980a) 'Season of Life - Season of Death', Social Science and Medicine, 13D, 227-232

Kevan, S.M. (1980b) 'Perspectives on Season of Suicide: A Review', Social Science and Medicine, 14D(4), 369-378

King, P.E. (1979) 'Problems of Spatial Analysis in Geographical Epidemiology', Social Science and Medicine, 13D, 249-252

Kitagawa, E. and Hauser, P. (1973) Differential Mortality in the United States: A Study in Socioeconomic Epidemiology, Harvard University Press, Cambridge, Mass

Kloos, N. and Thompson, J. (1979) 'Schistosomiasis in Africa: An Ecological Perspective', Journal of Tropical Geography, 48, 31-46

Knox, P.L. (1981) 'Convergence and Divergence in Regional Patterns of Infant Mortality in the United Kingdom from 1949-51 to 1970-71', Social Science and Medicine, 15D, 3, 323-328

McCracken, K.W.J. (1981) 'Analysing Geographical Variations in Mortality: Age-Specific and Summary Measures', Area, 13(3), 203-210

MacDonald, E.J. (1976) 'Demographic Variation in Cancer in Relation to Industrial and Environmental Influence', Environmental Health Perspectives, 17, 153-166

McGlashan, N.D. (1977) 'Spatial Variations in Cause-Specific Mortality in Australia', in N.D. McGlashan (ed.), Studies in Australian Mortality, University of Tasmania, Environmental Studies Occasional Paper 4, Hobart, 1-28

McGlashan, N.D. (1980) 'The Social Correlates of Alcohol-Related Mortality in Tasmania, 1971-1978', Social Science and Medicine, 14D, 181-203

McGlashan, N.D. and Grice, A.C. (1983) 'Sudden Infant Death and Seasonality in Tasmania, 1970-1976', Social Science and Medicine, 17(13), 885-888

Meade, M.S. (1978) 'Community Health and Changing Hazards in a Voluntary Agricultural Resettlement', Social Science and Medicine, 12D, 95-102

Meade, M.S. (1983) 'Cardiovascular Disease in Savannah, Georgia', in N.D. McGlashan, and J.R. Blunden (eds.), Geographical Aspects of Health, Academic Press, London, 175-196

Minowa, M. et al. (1981) 'Geographical Distribution of Lung Cancer Mortality and Environmental Factors in Japan', Social Science and Medicine, 15D(1), 225-231

Moens, G.F.G. (1984) 'Some Aspects of the Geographical Mortality Pattern of the Brussels Population in 1970', Social Science and Medicine, 18(1), 59-62

Mohan, J.and Rhind, D. (1983) 'Linking and Analysing Census and Medical Data: Mortality Experience in County Durham, 1969-1977', Working Paper No.21, Census Research Unit, Department of Geography, University of Durham

Mukerjee, A.B. (1980) 'The Disease Ecology of a Small cul de Sac: Chandigarh Dun', Social Science and Medicine, 14D, 331-337

Murray, M. N. (1967) 'The Geography of Death in the United States and the United Kingdom', Annals Association of American Geographers, 57, 301-314

National Centre for Health Statistics (1980) Health United States 1979, US Government Printing Office, Washington

National Centre for Health Statistics (1983) Health United States 1982, US Government Printing Office, Washington

Office of Health Economics (1979) Perinatal Mortality in Britain: A Question of Class,', OHE Briefing Paper No.10, Luton

Office of Population Censuses and Surveys (1978) Occupational Mortality, 1970-1972, DS Series No. 1, OPCS, London

Office of Population Censuses and Surveys (1980-82) Infant and Perinatal Monitor, 1979, 1980, 1981

Office of Population Censuses and Surveys (1982) Perinatal Mortality and Infant Mortality by Birthweight, 1981

Office of Population Censuses and Surveys (1983) Cigarette Smoking 1972-82, GHS 83/3

Paulozzi, L. (1981) 'The Seasonality of Mortality in Alaska', Social Science and Medicine, 15D, 335-339

Picheral, H. (1976) Espace et Sante, Geographie medicale du Midi de la France, Montpellier

Puffer, R.W. and Griffith, G.W. (1967) Patterns of Urban Mortality, World Health Organisation, Washington

Pyle, G.F. (1971) Heart Disease, Cancer and Stroke in Chicago, Department of Geography Research Paper No. 134, University of Chicago

Pyle, G.F. and Rees, P.H. (1971) 'Modelling Patterns of Death and Disease in Chicago', Economic Geography, 47, 475-488

Reid, J. (1982) Black America in the 1980s, Population Reference Bureau vol. 37(4), Washington

Robinson, V.B. (1978) 'Modelling of Spatial Variations in Heart Disease Mortality: Implications of the Variable Subset Selection Process', Social Science and Medicine, 12D, 165-172

Sato, T. et al. (1981) 'Geographical Studies of Seasonality in Cancer of the Stomach', Social Science and Medicine, 15D, 389-394

Schuman, S. (1972) 'Patterns of Urban Heat-Wave Deaths and Implications for Prevention: Data from New York and St Louis during July 1966', Environmental Research, 5, 59-75

Selya, R.M. (1980) 'Deaths Due to Accidents in Taiwan: A Possible Indicator of Development', Social Science and Medicine, 14D(4), 361-367

Shannon, G.W. and Dever, G.E.A. (1974) Health Care Delivery: Spatial Perspectives, McGraw-Hill, New York

Sindiga, I. (1984) 'Sleeping Sickness in Maasailand', Social Science and Medicine, 18(2), 183-187

Smith, D.M. (1982) 'Geographical Perspectives on Health and Health Care', in J. Cornwell, et al. (eds.), Contemporary Perspectives on Health and Health Care, Occasional Paper No. 20, Dept of Geography, Queen Mary College, London, 1-11

Stimson, R.J. (1982) The Australian City: A Welfare Geography, Longman Cheshire, Melbourne

Takahashi, E. (1981) 'Geographic Distribution of Cerebrovascular Disease and Environmental Factors in Japan', Social Science and Medicine, 15D(1), 163-172

Tamarisho, H. et al. (1981) 'Geographical Distribution of Cerebrovascular Diseases in Japan 1969-1974', Social Science and Medicine, 15D(1), 173-186

Thomson, N. (1982) 'Patterns of Aboriginal Mortality', Unpublished Paper Presented to Australian Population Conference, Canberra

United Nations (1982) Population of Papua - New Guinea, UNO, New York

Van Poppel, F.W.A. (1981) 'Regional Mortality Differences in Western Europe: A Review of the Situation in the Seventies', Social Science and Medicine, 15D(3), 341-352

Weil, C. (1981) 'Health Problems Associated with Agricultural Colonisation in Latin America', Social Science and Medicine, 15D(4), 449-461

Wilson, M.G.A. (1972) 'A Note on Infant Death in Melbourne', Australian Paediatric Journal, 8, 61-71

Wilson, M.G.A. (1978) 'The Geographical Analysis of Small Area/Population Death Rates: A Methodological Problem', Australian Geographical Studies, 16(2), 149-160

Wilson, M.G.A. (1979) 'Infant Death in Metropolitan Australia, 1970-1973', Canadian Studies in Population, 6, 127-142

Wood, D.R. (1982) 'The Spatial Distribution of Infant and Perinatal Mortality in South Australia and the Influence of Social, Economic and Demographic Factors, 1970-1981', unpublished BA(Hons) Thesis, Flinders University of South Australia

Woods, R. (1979) Population Analysis in Geography, Longmans, London
Woods, R. (1984) 'Spatial Demography', in J.I. Clarke, (ed.), Geography and Population: Approaches and Applications, Pergamon, London, 43-50

Chapter Five

GOVERNMENT POPULATION POLICIES

I. Thomas

INTRODUCTION

When the United Nations Department of Economic and Social Affairs produced its report on "The World Population Situation in 1970" (UN, 1971) it observed that all governments have policies, legislation and programmes which affect population growth and distribution. It then went on to assert: "However, such measures represent national population policy only when implemented for the purpose of altering the natural course of population movements." (p.67). This view was to be modified considerably over the next five years and in ways which brought the subject matter of population policy and population geography much closer.

The change in official attitudes was already occurring while the UN report was being published. In Britain, for instance, Sir Solly Zuckerman - then Chief Scientific Adviser to the Government - in a memorandum on "Population growth in the United Kingdom" to the 1970-71 Parliamentary Select Committee on Science and Technology (HMSO, 1971:35-7), noted three main aspects of population change in Britain in which the government had a particular interest. They were policies which influence population movement, the effects of population growth and movement, and family planning and population growth. The first of these, government policies likely to affect movements, were of two sorts: "regional policies" designed to improve conditions in those areas designated development and intermediate areas, and "dispersal policies" whose aim was to reduce congestion in the major urban areas. The second category covered those areas of national life for which the government has a planning responsibility and therefore needs to take account of the likely consequences of population change. This included the demand for transport facilities, the nature of housing programmes, the planning of education services, the training of skilled workers, and the regional allocation of investment. Finally, the provision of family planning services as part of the National Health Service since the 1967 Act made this an important part of the government's concern.

His statement is of interest in the context of this book for several reasons. First, Britain was not alone among developed nations in reviewing the nature and desirability of government policies designed to affect population (Berelson, 1974; and Table 5.1).

Secondly, it represents also the realisation that national population policy should concern itself not only with government activities which directly affect - and aim to affect - population growth, movement, and distribution but two other areas as well. Government activity indirectly affects population in a great variety of ways. But also, population has an internal dynamic whose effects are of importance for a range of government activities. The Zuckerman memorandum gives explicit recognition to the importance of each of these interactions and argues that the government should have a coherent population policy which includes them. This broad interpretation of the nature of population policy was repeated at the first World Population Conference (Mauldin et al., 1974) where the developing nations in particular emphasised "development" as their prime concern and gave attention to population insofar as it influenced the possibilities for achieving desired change (Wolfson, 1978). It was also examined

Table 5.1: Inquiries into Population Initiated by the Governments of Developed Nations, 1961-72

Israel	(1962)
Canada	(1967)
Israel	(1968)
Greece	(1968)
Japan	(1969)
Australia	(1970)
Britain	(1970)
France	(1970)
Hungary	(1970)
United States	(1970)
Argentina	(1971)
Britain	(1971)
Bulgaria	(1971)
Czechoslovakia	(1971)
France	(1971)
Romania	(1971)
Netherlands	(1972)

Source: Berelson, 1974a, pp.22-23, and Berelson, 1974b, 776-777.

systematically by Robinson (1975, p.2) who advocated this approach and contrasted it with what he characterised as a narrow conception of population policy concerned primarily with the "organised effort by government to affect population size, growth, distribution, or structure", or the even more restricted view of population policy as family planning programmes designed to lessen rates of population increase in low income countries. The general acceptance of this view has been followed by the adoption of population policies in a number of countries which formerly rejected them (Table 5.2 and Figure 5.1).

Thirdly, the Chief Scientific Adviser's memorandum also indicates quite clearly that many features of national life regularly investigated by geographers are central to the consideration of government population policy. This chapter examines the evolution and concerns of recent British population policy and some of the research undertaken following the government's recognition that the specification of policy is a complex and difficult matter. Then, the contribution of geographers to the analysis of the components of population policy is discussed with reference to work done in both Britain and in other countries.

Figure 5.2 provides a schematic framework linking policy, theory, and data collection via the related activities of problem identification, legislation, implementation, and monitoring and evaluation. National population policies bring together views on the problems of a nation and the determinants of those problems. This involves the perceptions of some group within the nation as to a set of significant national problems which can be ameliorated or eliminated by government intervention. It also involves some supposed understanding of sequences of cause and effect in which characteristics of population play an important part. The American demographer Bogue (1974) discussing population programmes with Asian policy makers in Bangkok echoed Keynes when he remarked:

> "Although he usually is not aware of it, each time a family planning administrator makes an important policy decision, he is making practical use (or abuse) of theories about human behaviour. A policy is proposed action to be taken in order to cause people to behave in a particular desired way. The reasons which the administrator gives for expecting his actions to have the desired effect are based upon his beliefs about why people behave as they do - in other words, upon an identifiable and formulable theory of human behaviour." (p.82, Bogue, 1974).

Government population policies reflect the views of members of a government but also the influence of internal and external

Table 5.2: Changes in the Commitment to Family Planning Programmes among Low and Middle Income Nations, 1972-1982

Index:	World Bank Income Groups:					
Degree of	Low Income		Lower Middle Income		Upper Middle Income	
Support	1972	1982	1972	1982	1972	1982
A. V. Strong	1(a)	1(a)	0	0	2(b)	2(b)
B. Strong	2	2	2	3	1	4
C. Moderate	1	3	7	7	4	3
D. Weak	3	5	6	11	1	2
E. V. Weak	25	21	21	15	6	3
Changed Commitment (Index change 1972-78: + towards A, - towards E)						
+1	5		11		3	
+2	1(c)		0		1(d)	
+3	0		0		1(e)	
-1	1		1		0	
-2	0		1(f)		0	
FP Cancelled	2(g)		1(h)		1(i)	

(a) China, (b) Rep. of Korea and Singapore, (c) Bangladesh, (d) Brazil, (e) Mexico, (f) Costa Rica, (g) Kampuchea (1977) and Lao PDR (1976), (h) Bolivia (1976), (i) Chile (1979).

Source: derived from World Bank, 1984.

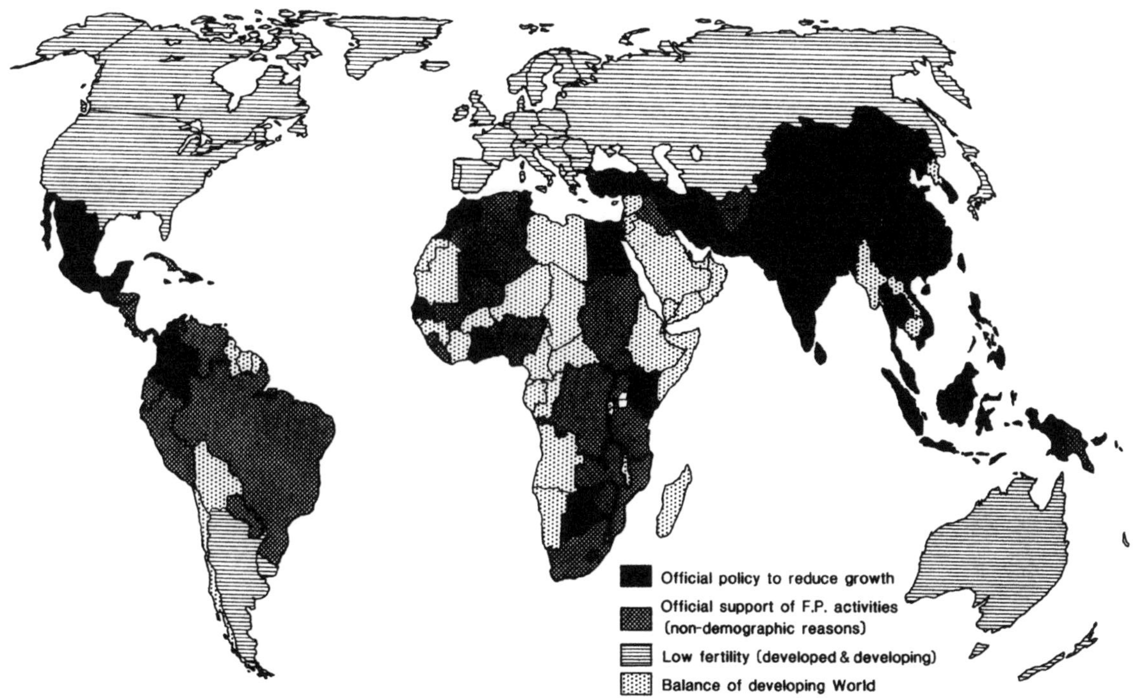

Fig. 5.1 World map of government family planning and population reduction programmes, 1978. Source: D. Nortman & E.Hofstatter, Population and family planning programmes: A compendium of data through 1978. 10th Edition New York: Population Council, 1980.

pressure groups (Godwin, 1975). It is probably true to say that internal pressure groups are most influential in industrialised nations, but international agencies both governmental (United Nations agencies and the World Bank, for instance) and non-governmental (the churches, Ford & Rockefeller Foundations, The Population Council, etc.) exert a major influence on the prevailing government population policy in many less developed countries. But the views of these interest-groups are themselves moulded by the findings of research and data-gathering, as well as by political considerations. Governments organise their own research enquiries and sponsor that of scholars, they also conduct monitoring and evaluation studies of programmes in the course of implementation. The work of Brass, Caldwell, Coale and Hoover, Bongaarts, Easterlin, Mabogunje and Todaro for instance, great contributors to scholarly research on population and its application to the practical affairs of government, feeds from this process and feeds into it. Theories may be divided between those which treat areas or the tendencies in great subpopulations (classes, urban dwellers, the educated, etc.), and those which are about individual behaviour and the factors which influence individuals. Fertility behaviour and migration arise from individual decisions, and many of the failures in population programmes have resulted from misunderstanding of the link between personal action and what governments may do to influence it. This essay seeks to examine components of the research-policy cycle, rather than to provide either a description of country programmes or a more direct account of the geographical concepts and techniques which might have relevance for formulating or evaluating population programmes. However, it consistently argues that whereas contributions from other disciplines - demography, economics, and sociology - have emphasised the factors which are associated with population dynamics, geographers have an intellectual responsibility to examine and account for the regional variation in the force of these factors and the results of their operation.

THE UK CASE: THE NATURE AND ORIGIN OF NATIONAL POPULATION POLICIES

Population Trends, the Royal Commissions, and the "Implicit Policy"

The UK Select Committee Report of 1971 (HMSO, 1971) was the result of meetings held to investigate the consequences of population growth in the United Kingdom. The previous two to three decades had been marked by an unexpected surge in population growth (OPCS, 1978). The population of the United Kingdom which was 46.0 million in 1931, had grown to 50.6 million by 1951 and 53.0 million by 1961. The 1971 census recorded a total of 55.7 million (HMSO, 1973, p.75). The two most recent large scale appraisals of the population of the United Kingdom, the 1940 Royal Commission on the Distribution of Industrial Population (Barlow Report), and the 1944 Royal Commission on Population (HMSO, 1949) had been initiated in wartime and after a period of concern of a different sort. This was anxiety about, first, the high levels of unemployment in the 1920s and 1930s and its uneven regional incidence and, secondly, the low level of fertility reached in Britain during the same period (Glass, 1936, 1940). Whereas the examination of population distribution was followed, in the postwar years, by powerful and influential legislation - in particular the Distribution of Industry Act of 1945, and the Town and Country Planning Act of 1947 (Armstrong and Taylor, 1978), the Royal Commission report on population growth was overtaken, as it were, by demographic events. By the late 1940s it was clear the population was growing vigorously, and subsequent surveys and censuses showed this to be more than an ephemeral postwar baby-boom. The legislation which emerged much later - the 1967 National Health (Family Planning) Act, and the Abortion Act - was antinatalist in tendency, though basically humanitarian in motivation. The immigration legislation of 1962, 1968 (Commonwealth Immigrants Acts) and 1971 (Immigration Act) also imposed controls which limited population increase (Jones, 1977).

British Government interest in population change by 1971 was affected by two considerations. First, by concern about population growth in the world at large and its implications for resource availability, trade and international stability. Secondly, by concern about the possible effects of the recent - and continuing - population increase at home. Britain, like all other west European countries still had low levels of fertility and population growth by international standards (see Figure 5.3), or by those of the nineteenth century, but there was an enhanced awareness of the implications of any but zero growth (Davies, 1967). Combined with welfare considerations it resulted in the emergence of a rather powerful set of policies which dampened population growth via both natural increase and net migration. Unlike France, where there has remained a strong preference for pronatalist policy (Spengler, 1938; Sauvy, 1974, pp.277-302; Dyer, 1978; Giscard D'Estaing, 1979), the British government approach to direct intervention to influence fertility behaviour had always been cautious and equivocal. In the late 1930s it was mildly pronatalist but by the late 1960s it had veered towards the antinatalist. Prior to the 1967 family planning legislation, official support for family planning services was contained only in an enabling memorandum from the Ministry of Health in 1930 which allowed Health authorities to provide information on birth control to married couples who, on medical grounds, needed it (Medwar & Pyke, 1971, pp. 56-7). After the 1967 National Health (Family Planning) Act, local authorities were empowered to give contraceptive advice without regard to marital status, and for social as well as medical reasons. The 1967 Abortion Act legalised abortion on grounds other than the preservation of the woman's life. It replaced, for the first time, the Offences Against the Persons Act of 1861 by which abortion, in any circumstances, was a crime attracting a maximum penalty of life imprisonment. Whereas these Acts were enabling, those introduced to influence international movements were more directive.

In 1962 the Commonwealth Immigrants Act introduced employment vouchers and subsequently country quotas, the 1968 Act of the same name extended these restrictions, and the 1971 Immigration Act imposed a unified system of control. Although there are severe limitations about the adequacy of data with which to gauge the impact of this legislation, Tables 5.3, 5.4 and 5.5 show something of the effect of these measures. The variable regional response to the availability of abortion facilities was commented on in the early years of operation of the new Act, but subsequent research on this and on the availability and use of the family planning facilities has tended to focus on national sample survey data and social class differentials to the exclusion of regional variations. The immigration legislation has imposed a tight control on the volume and geographical sources of new residents.

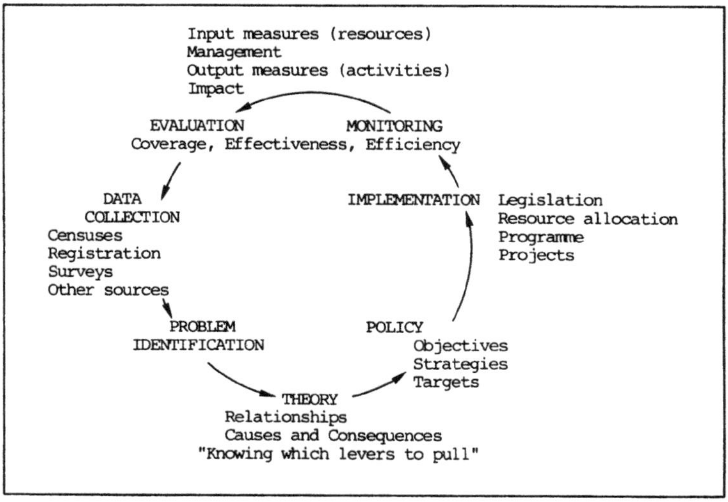

Fig. 5.2 Schematic links between research and government policy and programmes.

Table 5.3: Family Planning in England and Wales

a. Methods of contraception ever-used by women with a legitimate birth, 1967/8 to 1975. (Percentages, from sample surveys)

Method	1967/8	1970	1973	1975
Pill	28	40	65	74
IUD	3	4	8	13
Sheath	67	68	69	69
Withdrawal	46	45	45	31
None	7	6	3	1

Source: Cartwright 1976, Table 7, p.13; Cartwright 1978, Table 5, p.4.

b. Use of the family planning services, 1970 and 1975 (ever-married women under 41). (Percentages from sample surveys)

Category	1970	1975
Current users:		
GP user	14	19
Clinic user	9	16
Other service	1	2
Total current	24	37
Past users	27	31
Total ever-users	51	68
Never users	43	32
Not known	6	-
Total (surveyed)	100	100

Source: Bone, 1978, Table 2.1, p.6.
Note: Although the wider availability of family planning dates from the 1967 Act, subsequent changes under the National Health in the provision of free family planning services in clinics (1974) and by general practitioners (1975) are believed to be important (OPCS, 1978, p.29).

The Select Committee Recommendations

The Select Committee reached the general conclusion that "The government must act to prevent the consequences of population growth becoming intolerable for the every day conditions of life." (1971, p.x). Their principal recommendation was the creation of a Special Office charged with the duty of advising the government on population policy. In order to do this the Office was to coordinate and improve the

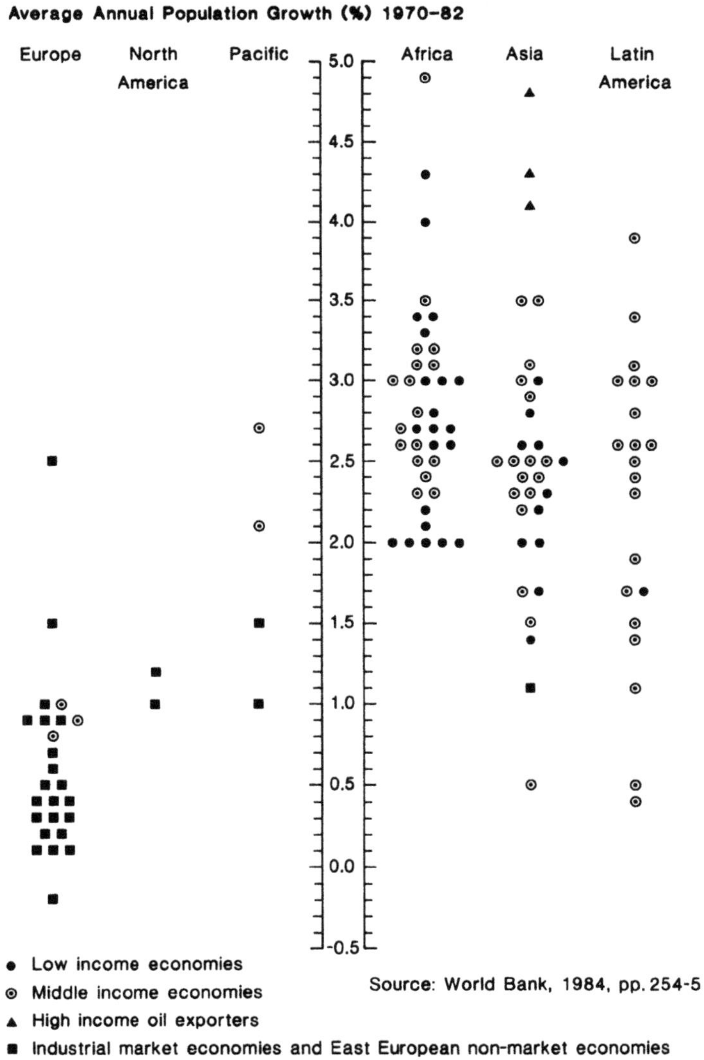

Fig. 5.3 National population growth 1970-1982, by continent and World Bank income groups.
Source: World Bank, 1984, pp.254-5).

Table 5.4: Abortion in England and Wales

a. Legally induced abortions (residents), England and Wales 1969-1983 (thousands)

1969	1971	1973	1975	1977	1979	1981	1983
50	95	111	106	103	121	129	127

Source: Population Trends 37, Autumn 1984, Table 23, p.50

b. Legally induced abortion rates (per thousand for women aged 15-49), by Standard Regions of England and Wales

Region	1968	1969	1970	1971	1972	1973
North	2.6	3.8	6.1	7.2	7.8	7.7
Yorks. & Humb.	1.4	2.7	4.8	6.2	7.1	7.7
North West	1.8	3.2	5.0	6.7	7.9	8.3
East Midlands	1.5	2.5	4.6	6.4	7.4	7.7
West Midlands	2.0	3.4	6.4	8.2	9.6	9.8
East Anglia	3.0	3.9	5.4	8.1	7.5	7.6
South East	4.5	6.3	8.9	10.8	12.6	12.6
South West	2.2	3.9	5.9	7.1	8.6	8.3
Wales	2.4	4.0	6.2	7.9	8.6	8.3
England & Wales	2.9	4.4	6.7	8.4	9.7	9.8

Source: Registrar General's Statistical Review of England and Wales for the Years 1968-73, Supplements on Abortion, Table 2(b).

study of UK and world population trends and their consequences, and to examine the implications for population policy of the plans of the various Departments of State (housing, water supplies, food, transport, fiscal policy, employment, education, health services). The Committee further recommended that the Special Office should be directly responsible to the Prime Minister who, it was suggested, should report annually to Parliament on its work.

The Ross Panel and its Recommendations

As a response to the findings of the Select Committee, the Government established a Population Panel in November 1971 in recognition of the fact that "The question of whether or not the Government should have a population policy or policies is complex and controversial." (HMSO, 1973:vii). It had

Table 5.5: Changes in the Volume of United Kingdom Immigration by Area of Origin 1966-1981

Year	Immigration All Countries	Old Commonwealth(a)	Other Commonwealth(b)
1966	219	36	79
1971	200	52	65
1976	191	40	67
1979	195	31	75
1981	153	20	56

Source: CSO, 1982, Table 1.13, p.19.
Notes: (a) Australia, Canada, New Zealand; (b) rest of the Commonwealth and Pakistan.

the brief "to assess the available evidence about the significance of population growth for both public affairs and private life in this country at present and in prospect: to make recommendations about the further work required, and how it should be conducted" (p.vii). The Panel concluded that whereas the population of Britain might increase by 20% by the end of the century or shortly thereafter, this increase could be accommodated and was not likely to be a significant contributory factor for the major anticipated national problems of the period. These problems were expected to arise from the necessity for structural change in industry, and a range of environmental and social pressures related more to levels of urbanisation, consumption, and alienation than to faster or slower rates of population increase. Nonetheless, the Panel concluded "that Britain would do better in future with a stationary rather than an increasing population." (p.6). This was based on the belief that most of the problems, though not caused in any major sense by population growth, would probably be easier to solve if the population was not growing, and that growth continuing into the 21st century would make it increasingly difficult to accommodate yet more people at that time. When it came to recommendations, the Panel took a restrained position on the grounds that, first, drastic action was not required, but - more significantly - also because "very little is known about the possible effects of different measures on population growth" (p.7). It felt able to make firm recommendations on measures to eliminate unwanted pregnancies: more comprehensive family planning services within the National Health Service, expansion of population education for the public and the health professions, and the elimination of inequalities in the provision of abortion facilities. Beyond this, the Panel called for more careful monitoring of population trends and study of the interaction of demographic and non-demographic conditions. In Government this was to be undertaken by a Minister with special responsibility for population matters, and a coordinating committee. These were to be provided with information from an expanded Office of Population Censuses and Surveys (OPCS), from recurrent stocktaking of population matters including mid-term censuses, and form the work of a new centre of demographic research within a university.

Developments since 1974

As indicated earlier, the Population Panel recommendations were made at a time when, yet again, there was a major change in the pace of British population growth. This removed much of the urgency with respect to internal policy considerations. An early casualty was the proposed 1976 mid-term census when, in the wake of the oil price crisis, the government was seeking to reduce expenditure. Nonetheless, for a short time there was a government minister responsible for coordinating population matters, and in the Overseas Development Administration a Population Bureau was maintained. The major developments, however, were not at the policy level nor in terms of fiscal programmes; they were broadly informational and in the supply of services. For example, the provision of family planning advice and supplies became a normal part of the public health service provision following the 1974 National Health Service reorganisation, and it has been recognised that the greater involvement of general practitioners in providing free family planning advice and supplies in addition to those available through clinics has lead to the use of family planning methods and services becoming "a fact of life for the vast majority of women" (Allen, 1981, p.115). Allen asserts that this has had an impact on the level of fertility during the 1970s, but Grebenik, who wrote the 1977 Demographic Review, is more circumspect (OPCS, 1978). The work of Ermisch (1983) suggests the critical forces are the participation rates of women in the labour force and the relative earnings of men and women: changes in motive are the crucial precedent to adoption of any form of birth control. The Demographic Review was one of a series of new publications which illustrate the second major outcome of the government review of population policy undertaken in the early part of the decade. There have been continuing developments in population education, the dissemination of population information, and in population research. The major contribution has come from government organisations: notably the Office of Population Censuses and Surveys, but also the research offices of other government bodies. The Department of Health and Social Security, for instance, monitored the family planning services (Bone, 1973; 1978) and the impact of an aging population (DHSS, 1981), and the Department of the

Environment continued to monitor the distribution of population and employment (DOE, 1971; DOE, 1976). These, and the publications of the OPCS - Population Projections, Population Trends, Social Trends, Regional Trends, and the various OPCS Monitors and Social Survey Reports, have been instrumental both in raising governmental and public awareness of the state of the nation's population, but also provided an important stimulus to further population research.

The first nationally representative surveys of birth control practice in Britain were organised by the Population Investigation Committee of the London School of Economics (Glass, 1976:p.x). These were carried out in 1959-60 and 1967-68. From 1967 the Office of Population Censuses and Surveys began a series of national sample investigations which have provided information on this topic. Glass commented in 1976, when five such studies had been carried out, "the fact that surveys of this kind are undertaken by the government represents a striking change in British official investigations" (p.xi). With the government taking responsibility for national surveys, Glass identified as areas for academic research: enquiries into the extent of infecundity and sub-fecundity, the causal connections between the employment of married women and fertility, and the means of establishing attitudes to and expectations of family size. However, it is relevant to note for this essay that the Langford/PIC study provided no information on regional variation. The OPCS Social Surveys and other government studies tend to focus on socio-economic differentials and derive their data from national sample surveys which are designed parsimoniously and do not permit regional disaggregation (Bone, 1978; Cartwright, 1978; and Dunnell, 1979). The OPCS Monitors on births, stillbirths, deaths, abortions, and other vital events have begun to give regional breakdowns, but their geographical analysis is a relatively new development.

Research funding for the universities was another of the Ross Report recommendations. No single new university centre has been created but funds have been provided directly and via the research councils to support training and research at the London School of Economics, in the Centre for Population Studies at the London School of Hygiene and Tropical Medicine (set up in 1974), the David Owen Centre for Population Growth Studies at Cardiff (also set up in 1974), and the Institute of Population Studies at Exeter. Geographical studies are not prominent at any of these centres. But at Durham (Census Research Unit), Newcastle (Centre for Urban and Regional Development Studies), Liverpool (African Population Mobility Project) and East Anglia (Overseas Development Group), to name just a few examples, funds have been provided to geographers for research and training in population geography or population census-related social and economic studies. One result of this cooperation between the government and the universities has been the series of annual conferences organised by the British Society of Population Studies - itself a product of the early 1970s - whose papers have been published by the Office of Population Censuses and Surveys (OPCS, 1982; OPCS, 1983).

COMPONENTS OF GOVERNMENT POPULATION POLICY AND SOME STUDIES BY GEOGRAPHERS

Although population geographers and other varieties of geographer have written at length about the distribution and movement of population, and have examined with considerable skill and sophistication the relationship of population to environmental and socio-economic conditions, they have not often directed their attentions specifically to the need for or the nature of population policies, and certainly not to the same extent or with the same impact as demographers and economists. Therefore, in examining the link between geographers and government population policies two approaches are adopted. First, the link is interpreted broadly rather than narrowly: it is argued, I believe with some justification, that geographers have played a not insignificant role in creating the climate for national population policies, in indicating the required components for policies, and in examining some of the possible implications of legislation and the consequences (intended and unintended) of its application. The review below is therefore presented in two parts: the first discusses direct studies of population policies, the second refers to work not directly focussed on specific population policies but which contributes to official and public awareness of either the need for policy or the consequences of policies. Secondly, some comparisons are made with the recent work of researchers in other disciplines and the way they have contributed to the formulation as well as the evaluation of national population programmes. This provides an opportunity to note recent population research which should be of interest to population geographers though not carried out by geographers.

Policy Studies: Policies designed to affect the rate of population growth

Geographers have not contributed to any great extent to the formulation, monitoring or evaluation of national population programmes designed to reduce population growth (Fuller, 1984) in sharp contrast to the volume of their work on programmes affecting the distribution and redistribution of population (Fuchs, 1984). Other social scientists - notably demographers, economists, and public health specialists - have been active in investigating and influencing population reduction policies (Bulatao and Lee, 1983; Easterlin, 1981; International Statistical Institute, 1981; Sirageldin et al., 1983; UN, 1973; Wolfson, 1978; World Bank, 1974, 1984). The arguments of political scientists have also figured prominently in the debate about the relevance of birth control programmes (Mamdani, 1972; Warwick, 1982). Harvey's paper of 1974 is one of the few contributions by geographers to this critical discussion. It is conducted at a high level of abstraction but reaches conclusions broadly consistent with the point of view adopted by many of the developing countries at the Bucharest World Population Conference (Mauldin et al., 1974), namely that the supposed problem of population was in fact one of the organisation of the economy and the disposal of resources. Interestingly, practice in many of these nations, and particularly in China, since 1974 has modified the interpretation of this position significantly: rapid population growth is now seen as something to be controlled because of its implications for national, social and economic development, but also in order to improve the health of mothers and children (World Bank, 1984).

Family Planning and Abortion

Blaikie's book on family planning in India (Blaikie, 1975) is the most extensive treatment of a government birth control programme available from a geographer. It is a strange reflection on the reticence of geographers that the recent review of population geography in India by Ghosal (Clarke, 1984) barely mentions the family planning programme of India. Blaikie's study provided a vigorous critique of many aspects of the management of the Indian programmes of the late 1960s and early 1970s together with clear and numerous recommendations for improvement. These messages were derived from an analysis of detailed field investigations, innovative cartographic presentation and statistical investigation. The theoretical underpinning was spatial diffusion, but the conclusions emphasised the need for geographical and cultural specificity and sensitivity in programme implementation. In this instance, the geographical perspective may have obscured - for practical purposes - the policy and programme relevance insofar as the family planning service was being used as the case with which diffusion theory was being investigated. The effect of the 1975-76 emergency on the population programme of India (Gwatkin, 1979) also served to lessen its chance of having an impact on programme design. Despite the excellence of its analysis, the study has not received the attention it warrants from those who design, implement and evaluate programmes in India (Dyson and Crook, 1984). Nortman's comparison of national development plan targets and achievements (Figure 5.4) demonstrates the continuing need for research into fertility determinants in India. Gwatkin drew attention to the probable significance of distance from Delhi as a factor in explaining the differential impact of the drive for more effective birth control during the emergency. More recently, the State contrasts in fertility level have invited speculation and explanation. Despite its comparative poverty Kerala has reached the lowest crude birth rate (Preston and Bhat, 1984), and this has been attributed to the combined effect of general improvements in health and education and redistributive policies (Zachariah and Kurup, 1982). The role of diffusion, the existence of urban bias, and the impact of distance on acceptance and use are recurring themes in the small amount of work by geographers on both family planning and abortion facilities (Coward, 1978; Fuller, 1973; Henry, 1978, 1982; Jones, 1977). As noted in the UK case study, national sample investigations of service availability and accessibility have been an outcome of British population policy, but there have been relatively few comparative and regional studies (Moseley, 1979; Bentham, 1984). The international significance of abortion, for instance, as a component of population control, and the regional variations in both its occurrence and acceptability to governments has been documented over the years by Tietze (1983) but there has been little in the way of regional causal analysis. As a second example, the recent work of Page and Lesthaeghe (1981); Lesthaeghe (1984); Bongaarts et al. (1984), on child-spacing in Africa demonstrates clearly the need for demographically informed studies of the cultural diversity even within nations, and how government population policies must be responsive to these variations and their sometimes rapid modification, as in the case of lactation practices and the operation of customary taboos on post-partum abstinence. These studies indicate the need for new emphases in existing programmes, and the possible constraints on efficiency, but there are rather few of them by geographers. This is despite a recognition among non-geographer researchers of both the problems and potential value of a real investigation (Duncan et al., 1961; Hermalin, 1975), and the usefulness of comparative studies (Watson and Lapham, 1975).

Mortality

There have, if anything, been even fewer geographical studies of the mortality component of national population policies. Most development policies in Third World countries will have either an explicit statement of their interest in reducing mortality, especially infant and child mortality, or this intention will be subsumed by the programmes for health, nutrition and water. Tanzania, for example, shortly after independence in 1961 included in its first five year plan a target to increase the expectation of life from the then current 35/40 years to 50 years by 1980 (Thomas, 1972).

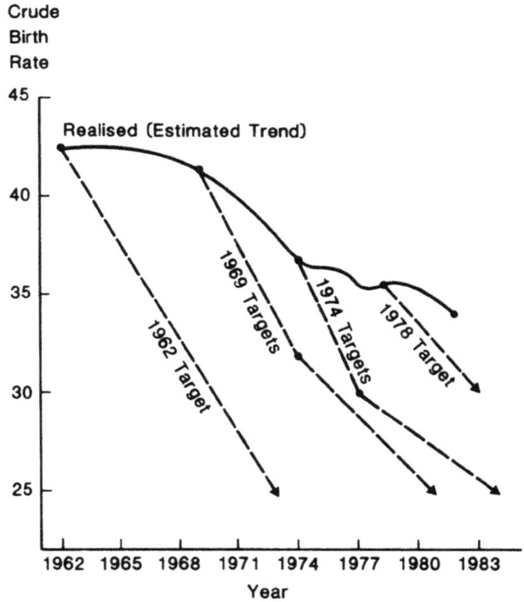

Fig. 5.4 Trend of crude birth rate in India 1962-82 and the targets in successive five-year development plans. Source: D.L. Nortman, 'India's new birth rate target: An analysis', Population and Development Review, 4.2 (June, 1978), p.281. The 1982 crude birth rate estimate is from World Bank, 1984, p.254).

Recently (Tanzania, 1981) the government has re-affirmed this commitment and set the target for the end of the century at 55 years. Curson (1984, pp.95-6) has rightly drawn attention to the technical problems of research on the spatial patterning of mortality, health and disease, but these are no more intractable than the problems faced by demographers in collecting and analysing mortality data in developing countries (Kpedekpo, 1981), and persistent effort in this direction has been richly rewarded (Preston and Bennett, 1983; UN, 1983). The methods developed are now being applied to identify regional and social differentials, and to refine population targets and improve social welfare programmes (Flegg, 1982; Haines and Avery, 1982; Hobcraft et al., 1984; Martin et al., 1983). The geographical literature on social provision has begun to include international studies of the spatial inequalities of health care systems and the demographic indicators - including mortality measures - with which success may be judged (Thomas, 1983; Prothero and Gould, 1984; Phillips and Joseph, 1984). In developing countries the rapid change in mortality has been a major immediate cause of the high rate of overall population increase, and it is one of the dilemmas which government policy makers in those countries face: while trying to reduce the rate of population increase they are also seeking to reduce mortality.

Studies of regional, social and cause-specific mortality differentials in developed countries are more common (for some UK and Australian examples see: Chilvers, 1978; Howe, 1970, 1979; McCracken, 1981; McGlashan, 1977; Wilson, 1979), but they relate more closely to health and environmental policies and programmes than directly to population programmes. The recent concern in Britain about various forms of infant mortality, and in the USSR about the trend in infant death rates are perhaps exceptions (Festbach, 1984).

Population Composition: some consequences of policies and trends

An interesting subject area which, in Britain at least, involves both the effects of natural increase and the results of selective migration, is the concentration of elderly people in particular areas of the country. This is clearly of direct concern for social policy in general, and for area health and local authority planning in particular (Warnes, 1982; Warnes and Law, 1984). A second example, also from Britain, is the political and social implications of the selective immigration policy and its effect on both population growth and ethnic concentration and segregation (Peach, 1968; Anwar, 1979).

Policy Studies: Policies designed to affect population distribution and redistribution

Geographers have made a much greater contribution to debates on the whole range of policies and programmes which may be described generally as distribution, redistribution and migration policies. This applies to both developed and developing nations. No attempt is made here to provide a systematic review even of the work on population redistribution and immigration in Britain. There have been several recent reviews of migration theory by geographers (White and Woods, 1980; Jones, 1981; Woods, 1982), and these and others have also examined redistribution policies (Mabogunje, 1981; Gilbert, 1982; Fuchs, 1984). None have attempted the comprehensive theorising of Zelinsky (1971) who formulated a mobility transition theory to parallel that of the demographic transition. The critique of the latter built up over nearly forty years by demographers, economists and sociologists has had a massive impact upon the thinking of aid agencies and governments, and still inspires basic and applied research (Caldwell, 1982), but the work of geographers on redistribution and migration - though influential - has been more fragmented. In the last few years, a series of symposia organised by the Population Geography Commission of the International Geographical Union has been devoted to population redistributions of all sorts (Clarke and Kawabi, 1980; Webb et al., 1981; Clarke and Kosinski, 1982; Clarke et al., forthcoming; Kosinski and Elahi, forthcoming). Since population distribution policy is as much about programmes for areas as it is about people (Wander, 1969), several compilations on regional development policy and planning overseas have included discussion of agricultural resettlement and other forms of rural-rural migration, rural-urban migration, intra-urban relocations, growth centre and secondary city policies, and counter-urbanisation (Rondinelli and Ruddle, 1978; Stohr and Taylor, 1981; Gilbert and Gugler, 1982; Lea and Chaudhri, 1983). Similar topics have been included in the work on rural geography and on urban and regional development in Britain (Goddard and Champion, 1983; House, 1982; Pacione, 1983). The dividing line between population geography and various other sub-disciplines both within geography and in other social sciences becomes blurred at this point. This is to be expected when dealing with policy issues: an initial focus on settlement policy, health, transport, agriculture, urban development, etc., will lead to consideration of population dynamics. Equally, a starting point with population distribution or movement will inevitably lead to these cognate subjects. This is the implication of the view of population policy which emerged after the Bucharest Conference. This comprehensiveness has lead some analysts (Gilbert, 1982, is a recent example) to suggest that migration policies will succeed only when whole social and economic systems are redesigned. Earlier, Berry (1973) had argued that the only states in which migration control policies work are authoritarian: he had in mind the Republic of South Africa and Israel. Gilbert takes as his examples Tanzania and China. No doubt debates at the level of grand political theory will continue, but there is still a need for detailed investigation of the association of specific legislation or institutional arrangements and the rate, composition and direction of migration. As an example, there has been useful work on the relation of education policy and migration (Barnum and Sabot, 1976; Bell, 1980), drawing on the concepts of place utility (Wolpert, 1965) and expected income differentials (Rempel and Todaro, 1972). Bell, working in Uganda, has demonstrated the importance of early migratory experience, much of it resulting from a school placements policy, on subsequent place utilities. Sabot's work in Tanzania lead him to conclude that the content of rural education encouraged unrealistic income expectations and stimulated migration to the urban areas, but that economic policies as well as curriculum reform were now needed to stem the flow.

The range of work on Africa by geographers provides an opportunity to illustrate the types of distribution and redistribution policies. Adepoju (1982) notes four policy types: rural development, resettlement schemes, youth programmes, and growth poles/administrative decentralisation. If we add to these the types of programmes devised to deal with unemployment in Third World cities or the supposed problem of excessive rural to urban migration (Todaro, 1976; Gilbert, 1982; World Bank, 1984), and add to the specifically rural and urban area policies those with a national scope which aim to alter the internal terms of trade between rural and urban areas, the range of policy-related subjects becomes wide (Table 5.6), and yet many of them have been the subject of geographical enquiry as the illustrative references in the table demonstrate.

The objectives of these programmes may be identified as:

1. to encourage an adequate supply of wage labour at particular locations (rural or urban);
2. to prevent people leaving rural regions - rural areas and small towns;
3. to encourage agricultural area extension (the spread of settlement);
4. to increase agricultural productivity (by intensification of use or by mechanisation);
5. to free land for other uses - rural (forests, game, reservoirs, large-scale agricultural production, mining) and urban (squatter resettlement for commercial/residential/industrial/transport development);
6. to prevent people entering towns;
7. to prevent people staying in towns;
8. to provide employment/housing etc. in towns (thereby encouraging migration); and
9. to produce a supposedly more efficient settlement pattern (judged by economic-spatial, or political criteria - both administrative and control).

Underlying the objectives and the chosen means are those "theories" to which Bogue referred, in this case: theories of capitalist development, theories of spatial efficiency, microeconomic theories of individual response to incentives, theories of search behaviour, and theories of social behaviour and control. Weiner (1975) has exposed some of the assumptions and purposes behind government migration policies and, in particular, drawn attention to the possible reluctance of governments to accept, or at least make public, the demographic intentions of apparently unrelated legislation, for example land use zoning designed to produce ethnic segregation.

As indicated earlier, geographers have demonstrated the need for policies by describing population distribution and population characteristics but also by relating the conditions of the population to environmental and socio-economic conditions. They have described causes of movements, the characteristics of movement and the selective nature of migration. They have described the consequences of movement for the individuals involved and for the societies left behind and those at the destination. In these ways they have contributed to the formulation and evaluation of redistributive policies and programmes to an extent which is unmatched in other areas of population policy.

Geographical Studies which indirectly bear upon policy. Regional studies and atlases

In a less tangible fashion, much of the geographical literature which is not directly addressing itself to policy formulation and evaluation nonetheless contributes to public awareness of population issues. This applies particularly to the substantive work describing and analysing the population distribution and composition, migration, urbanisation, and population-resource relations for specific areas, but it also applies to many thematic studies concerned - among other topics - with agriculture, transportation, and housing. There is an embarrassment of riches and only selected examples, some of them now classics, are given for Africa (Barbour and Prothero, 1961; Hance, 1970), Asia (Clarke and Fisher, 1971; Farmer, 1977; Trewartha, 1965), and Latin America (James, 1950; Santos, 1979; Gilbert, 1982). In Britain, the cartographic work of Hunt et al. (1968), Howe (1970), Shepherd (1974), Dewdney et al. (1980), Champion (1983) and a whole range of regional analysts of the decennial censuses is illustrative of this contribution.

Methodological Studies

Geographers have also contributed extensively to the examination of population data and the methods for its collection and tabulation (Denham, 1980; Nag, 1984; Rees and Wilson, 1977; Rhind, 1983). Whereas this may not seem directly related to government population programmes, the pressure for the collection of certain types of information and the public availability of data are often closely connected with the formulation of social and economic policy and programmes. It also permits the evaluation of the impact of policies. Geographers in Britain have figured prominently in the debates about migration and ethnic questions in the censuses and have contributed to the development of techniques for using grid-square and postcode-located information. Their work on employment, the availability of household amenities, and

Table 5.6: A Classification of Internal Migration and Redistributive Programmes with Selected Examples from Africa

Area Focus	Intended Effect on Movement: A. Increase/Facilitate	B. Diminish/Eliminate
I. Rural		
	1. Frontier settlement - river basin - cropland extension (Eastern Africa: Allan, 1965; McMaster, 1961; Egero, 1974) 2. Labour migration - direct: quotas recruitment - indirect: poll/head tax education (Southern Africa: Prothero, 1974; Coleman, 1979; Swindell, 1979) 3. Agricultural resettlement - land reform - irrigation schemes (and reservoirs) - river basin schemes - resettling reserves - disease-related - creation of resource reserves (forest, game) (Kenya: Odingo, 1971; Sudan: Davies, 1964, Khogali, 1982; Uganda: Kabera, 1982) 4. Villagisation (Algeria: Sutton & Lawless, 1978; Botswana: Silitshena, 1982; Tanzania: Maro & Mlay, 1982; Thomas, 1982)	1. Rural development - agricultural production - rural production infrastructure - rural social infrastructure - rural employment 2. Integrated regional development programmes (Todaro, 1971; Mabogunje & Faniran, 1977; Belshaw, 1982, Benneh, 1984)
II Urban		
	1. Labour recruitment/job creation 2. Urban housing provision 3. New Capitals 4. Secondary towns/growth centres - employment deconcentration - social service distribution - industrial dispersal - administrative dispersal 5. "Service centre" resettlement 6. Repatriate urban unemployed (West Africa: Adejuyigbe, 1970; Kudiabor, 1977; Mabogunje, 1978; Eastern Africa: Funnell, 1976; Davies, 1981; Hoyle, 1979; and general: Todaro 1976; O'Connor, 1983)	1. In-migration control - direct: pass law work permit schemes cash guarantees until employment - indirect: ration cards (food/clothes) housing controls (Southern Africa: Fair & Davis, 1976; Smith, 1982)

III National		
	1. Disaster responses - relocate population - refugee reception 2. International migrants - encourage (recruitment, subsides) - repatriate (W. Africa: Peil, 1971; Uganda: Twaddle, 1975; refugees: Rogge, 1982)	1. Moral exhortation: Education, political 2. Fiscal: - remove factor-price distortions - wage subsidies - income policies 3. Immigration & emigration restriction - direct: visa restrictions - indirect: foreign exchange restrictions 4. Disaster responses - flood control - rural food storage & free distribution - food for work programmes (Tanzania: Sabot, 1979; McCall, 198 ; general: Todaro, 1971; Abbo, 1982)

commuter catchments, to name but a few, have been instrumental in both keeping questions in the national censuses and monitoring the effects of social policies. Work elsewhere by population geographers has also been in cooperation with census organisations to ensure adequate data collection and presentation, and has demonstrated the continued existence of regional disparities which warrant new government initiatives or the modification of existing programmes where these are failing to achieve their targets.

CONCLUSION

Governments in both the developed and developing countries have taken more explicit account of population change during the last two decades. In both types of area concern has been expressed about population growth and population distribution, although the origin of that concern and the responses differ with national circumstances. When the elucidation of population policies has been followed by legislation and the creation of programmes designed to implement policy - as with family planning and population redistribution programmes - the outcome has often been disappointing. Social scientists have been active in describing these interventions and seeking to explain the varying degrees of impact. Geographers have contributed to policy related studies of distribution and migration, but much less to those concerned with rates of natural change. There are opportunities here for useful comparative analytical work, particularly as governments seek increasingly to understand intra-national variations or eliminate regional inequalities. In addition, a further development of national population policy during the past decade has been greater attention to the interrelation of population distribution and change with general social, economic and environmental conditions and processes. This is much closer to the traditional concerns of geographers who have a responsibility to demonstrate the value of their approach.

REFERENCES

Addo, N.O. (1982) 'Government Induced Transfers of Foreign Nationals', in Clarke and Kosinski, ch.4, pp. 31-38

Adejuyigbe, O. (1970) 'The Case for a New Federal Capital in Nigeria', Journal of Modern African Studies, 8, 301-6

Adepoju, A. (1982) 'Population Redistribution: A Review of Government Policies' in Clarke & Kosinski, Ch.8, pp. 58-65

Allan, W. (1965) The African Husbandman Edinburgh: Oliver & Boyd

Allen, I. (1981) Family Planning, Sterilisation and Abortion Services, Study No. 595, Policy Studies Institute, London

Anwar, M. (1979) The Myth of Return, London

Armstrong, H.& J. Taylor (1978) Regional Economic Policy and Its Analysis Oxford: Philip Allan

Barbour, K.M. and R.M. Prothero (1961) Essays on African Population. London: Routledge and Kegan Paul

Barlow Report (1940) Royal Commission on the Distribution of the Industrial Population Cmnd.6153. London HMSO

Barnum & R.H. Sabot (1976) Migration, Education and Urban Surplus Labour: The Case of Tanzania. Paris: Development Centre, Organisation for Economic Cooperation and Development

Bell, M. (1980) 'Past Mobility and Spatial Preference for Migration in East Africa', Ch.6, in White and Woods, pp. 84-107

Belshaw, D.G.R. (1982) 'An Evaluation of Foreign Planning Assistance to Tanzania's Decentralised Regional Planning Programme', Applied Geography, 2, 291-302

Benneh, G. (1984) 'Planning and Implementation of Rural Development Projects in the Upper East Region of Ghana' in T. Craig (ed.), Technological Innovation: University Roles, London: The Association of Commonwealth Universities, pp. 127-34

Bentham, C.G. (1984) 'Mortality Rates in the More Rural Areas of England and Wales', Area, 16, 3(Sept.) pp. 219-226

Berelson, B. (1974a) 'World Population: Status Report 1974. A Guide for the Concerned Citizen' Reports on Population/Family Planning, 15 (Jan.), pp.1-47

Berelson, B. (1974b) Population Policies in Developed Countries, New York: McGraw Hill

Berry, B.J.L. (1973) The Human Consequences of Urbanisation. London: Macmillan

Blaikie, P.M. (1975) Family Planning in India: Diffusion and Policy. London: Edward Arnold

Bogue, D.J. (1974) Policy Implications of Theory and Research on Motivation and Induced Behaviour for Fertility and Family Planning. Ch.XII, Asian Population Studies Series No.26 Bangkok: UN.ESCAP

Bone, M. (1973) Family Planning Services in England and Wales. An Enquiry Carried Out On Behalf of the Department of Health and Social Security. London: HMSO (for OPCS Social Survey Division)

Bone, M. (1978) Family Planning Services: Changes and Effects. Social Survey Report (SS 1055) HMSO.

Bongaarts, J. and R.G. Potter (1983) Fertility, Biology, and Behaviour: An Analysis of the Proximate Determinants. New York: Academic Press

Bongaarts, J., Frank, O. & R. Lesthaeghe (1984) 'The Proximate Determinants of Fertility in Sub-Saharan Africa', Population and Development Review, 10, 3 (Sept.), pp. 511-37

Brass, W., Coale, A.J. et al. (1968) The Demography of Tropical Africa. Princeton: Princeton University Press

Bulatao, R.A. & R.D. Lee (eds.) (1983) Determinants of Fertility in Developing Countries. 2 vols. New York: Academic Press

Caldwell, J.C. (1982) The Theory of Fertility Decline. London: Academic Press

Cartwright, A. (1976) How Many Children? London: Routledge & Kegan Paul

Cartwright, A. (1978) Recent Trends in Family Building and in the Use of Contraception. Studies on Medical and Population Subjects No.34. London: HMSO

Casetti, E. & W.L. LI (1979) 'The Family Programme in Taiwan: Did it Make Any Difference?' Geographical Analysis, 11. pp. 395-403

Central Statistical Office (1973) Social Trends, No.4, 1973. London: HMSO

Central Statistical Office (1982) Social Trends, No.13, 1983 Edition. London: HMSO

Champion, A.G. (1983) England and Wales '81. Sheffield: The Geographical Association

Chilvers, C. (1978) 'Regional Mortality 1969-73', Population Trends, 11, pp.16-20

Clarke, J.I. (ed.) (1984) Geography and Population: Approaches and Applications. Oxford: Pergamon

Clarke, J.I. & W.B. Fisher (eds.) (1972) Populations of the Middle East and North Africa: A Geographical Approach. London: London University Press

Clarke, J.I. and H. Kawabi (eds.) (1980) 'Population Redistribution in Asia and the Pacific', Population Geography (Chandigarh), Special Number
Clarke, J.I., Khogali, M.M. & Kosinski, L.A. (eds.), (forthcoming) Population and Development Projects in Africa. Cambridge: Cambridge University Press
Clarke, J.I. & Kosinski, L.A. (eds.) (1982) Redistribution of Population in Africa. London: Heinemann
Coale, A.J. & E.M. Hoover (1958) Population Growth and Economic Development in Low-Income Countries. Princeton: Princeton University Press
Coleman, G. (1979) 'International Labour Migration from Malawi', Occasional Paper No.1, School of Development Studies, University of East Anglia, Norwich
Coward, J. (1978) 'Family Planning Clinics in The Republic of Ireland', Irish Geography, 11, pp.189-92
Curson, P. (1984) Geography, Epidemiology and Human Health, in Clarke, 1984
Davies, D.H. (1981) 'Towards an Urbanization Strategy for Zimbabe', Geojournal, Supplement 2, 73-84
Davies, H.J.R. (1964) 'An Agricultural Revolution in the African Tropics: The Development of Mechanised Agriculture in the Clay Plains of the Republic of the Sudan', Tijdschrift Voor Economische en Sociale Geografie, 55, 101-8
Davies, K. (1973) 'Zero Population Growth', Daedulus, Fall 1973, pp. 15-30
Denham, C. (1980) People of Britain. London OPCS
Denham, C. (1980) 'The Geography of the Census 1971 and 1981', Population Trends, 19, pp.6-12
Dept. Of The Environment (1976) British Cities: Urban Population and Employment Trends 1951-71. DOE Research Report No.10. London: HMSO
Dept. Of Health & Social Security (1981) Growing Older. Cmnd.8173. London:HMSO
Dept. Of The Environment (1971) Long Term Population Distribution in Great Britain - a Study. Dept. of the Environment. London:HMSO
Dept. Of The Environment (1976) British Cities: Urban Population and Employment Trends 1951-71. Research Report No.10. London: HMSO
D'Estaing, G. (1979) 'President Giscard d'Estaing on Fertility Decline in France', Population and Development Review, 5, 3 (Sept.), pp. 571-3
Dewdney, J. et al. (1980) People of Britain. London: OPCS
Duncan, O. Cuzzort, Duncan, R.& B. (1961) Statistical Geography. Glencoe, III.: Free Press
Dunnell, K. (1979) Family Formation, 1976. Social Survey Report (SS 1080). London: HMSO
Dyer, C. (1978) Population and Society in Twentieth Century France. Sevenoaks: Hodder and Stoughton

Dyson, T. & N. Crook (eds.) (1984) India's Demography: Essays on the Contemporary Population. New Delhi: South Asian Publishers

Easterlin, R.A. (ed.) (1980) Population and Economic Change in Developing Countries. Chicago, Ill. : Chicago Univ. Press

Egero, B. (1974) 'Migration and Economic Development South of Lake Victoria', Research Paper No. 32. Bureau of Resource Assessment and Land Use Planning, University of Dar es Salaam

Ermisch, J. (1983) The Political Economy of Demographic Change: Causes and Implications of Population Trends in Great Britain. London: Heinemann

Fair, T.J.D. & R.J. Davies (1976) 'Constrained Urbanisation: White South Africa and Black Africa Compared', in B.J.L. Berry (ed.): Urbanisation and Counter-Urbanisation. London: Sage

Farmer, B.H. (ed.) (1977) Green Revolution? London: Macmillan

Festbach, N. (1984) 'On Infant Mortality in the Soviet Union', Population and Development Review, 10, 1 (March), pp. 87-90

Flegg, A.T. (1982) 'Inequality of Income, Illiteracy and Medical Care as Determinants of Infant Mortality in Underdeveloped Countries', Population Studies, 36, 3 (Nov.), pp. 441-458

Fuller, G.A. (1973) 'The Diffusion of Illegal Abortion in Santiago de Chile: the Use of a Direction-Bias Model', Proceedings of the Association of American Geographers, 5, 71-74

Fuller, G.A. (1984) 'Population Geography and Family Planning', in J.I. Clarke (ed.), Geography and Population: Approaches and Applications. Oxford: Pergamon, pp. 103-109

Funnell, D.C. (1976) 'The Role of Small Service Centres in Regional and Rural Development with Special Reference to Eastern Africa', in A. Gilbert (ed.), Development Planning and Spatial Structure. London: Wiley

Gilbert, A. (ed.) (1982) Urbanisation in Contemporary Latin America: Critical Approaches to the Analysis of Urban Issues. Chichester: Wiley

Gilbert, A. (1982) 'Urban and Regional Systems: A Suitable Case for Treatment', in Gilbert, A. and J. Gugler, ch.8, pp. 162-197

Gilbert, A. and Gugler, J. (1982) Cities, Poverty, and Development: Urbanisation in the Third World. Oxford: Oxford University Press

Glass, D.V. (1936) The Struggle for Children. Oxford: Clarendon Press

Glass, D.V. (1940) Population Policies and Movements in Europe. Oxford: Clarendon Press

Glass, D.V. (1976) 'Preface, pp. ix-xi, in C. M. Langford (ed.), Birth Control Practice and Marital Fertility in Great Britain, London: Population Investigation Committee

Goddard & Champion (eds.) (1983) The Urban and Regional Transformation of Britain. London: Methuen

Godwin, R.K. (ed.), (1975) Comparative Policy Analysis. Lexington: D.C. Heath

Gwatkin, D.R. (1979) 'Political Will and Family Planning: The Implications of India's Emergency Experience', Population and Development Review, 5, 1 (March), pp. 29-59

HMSO (1940) Royal Commission on the Distribution of Industrial Population. (Barlow Report) Cmnd. 6153 London: HMSO

HMSO (1949) Report of the Royal Commission on Population. Cmnd 7695, London: HMSO

HMSO (1971) Population of the United Kingdom. First report from the Select Committee on Science and Technology Session 1970-71. London: HMSO

HMSO (1973) Report of the Population Panel. (Ross Report) Cmnd. 5258. London: HMSO

Haines, M.R. & R.C. Avery (1982) 'Differential Infant and Child Mortality in Costa Rica: 1968-1973' Population Studies, 36, 1 (March), pp. 31-44

Hance, W. (1970) Population, Migration, and Urbanisation in Africa, New York: Columbia University Press

Harvey, D. (1974) 'Population, Resources, and the Ideology of Science', Economic Geography, 50, 3 (July), pp. 256-77

Henry, N.F. (1978) 'The Diffusion of Abortion Facilities in the Northeastern United States, 1970-76, Social Science and Medicine, 12, pp. 7-12

Henry, N.F. (1982) 'Regional Dimensions of Abortion-Facility services. Professional Geographer, 34, 1, pp. 65-70

Hermalin, A.I. (1975) 'Regression Analysis of Area Data', in C. Chandrasekaran & A.I. Hermalin (eds.), Measuring the Effect of Family Planning Programs on Fertility. Dolhain: Ordina, Ch.8, pp. 245-99

Hobcraft, J.N. MacDonald, J.W. & Rutstein, S.O. (1984) 'Socio-Economic Factors in Infant and Child Mortality', Population Studies, 38, 2 (July), pp. 193-224

House, J.W. (ed.), (1982) The U.K. Space: Resources, Environment and the Future London: Weidenfeld and Nicolson

Howe, G.M. (1970) National Atlas of Disease Mortality in the United Kingdom. London: Nelson

Howe, G.M. (1979) 'Mortality From Selected Malignant Neoplasms in the British Isles: The Spatial Perspective', Geographical Journal 145, 401-415

Hunt, A.J. (ed.), (1968) 'Population Maps of the British Isles 1961', Transactions of Institute of British Geographers, 43 (April)

International Statistical Institute (1981) World Fertility Survey Conference: Report of Proceedings. London/Voorburg. ISI. 3 volumes

James, P. (1950) Latin America. New York: Odyssey Press

Jones, H.R. (1977) 'Metropolitan Dominance and Family Planning in Barbados', Social and Economic Studies, 26, pp.327-38

Jones, H.R. (1981) A Population Geography. London: Harper and Row

Jones, K. (1977) Immigration and Social Policy in Britain. London: Tavistock

Kabera, J.B. (1982) 'Rural Population Redistribution in Uganda since 1900', in Clarke & Kosinski, Ch. 26, pp. 192-201

Khogali, M.M. (1982) 'Western Sudanese Migrants to Kashm el-Girba Agricultural Region', in Clarke & Kosinski, Ch. 23, pp. 166-75

Kosinski, L.A. & K.M. Elahi (eds.), (forthcoming) Population Redistribution and Development in South Asia. Dordrecht: Reidel

Kpedekpo, G.M.K. (1981) 'Mortality', in ISI 1981, vol.2, pp. 473-502

Kudiabor, C.D.K. (1977) 'Urbanisation and Growth Pole Strategy for Regional Development in Ghana', in A.L. Mabogunje and A. Faniran (eds.), Regional Planning and National Development in Tropical Africa. Ibadan: Ibadan University Press

Lea, D.A.M. and D.P. Chaudhri (eds.) (1983) Rural Development and the State: Contradictions and Dilemmas in Developing Countries. London: Methuen

Lesthaeghe, R. (1984) 'Fertility and its Proximate Determinants in Sub-Saharan Africa: The Record of the 1960s and 1970s', Interuniversity Programme in Demography Working Paper 1984-2. Brussels: Vrije Universiteit

Mabogunje, A.L. (1978) 'Growth Poles and Growth Centres in the Regional Development of Nigeria', in A. Kuklinski (ed.), Regional Policies in Nigeria, India and Brazil. The Hague: Mouton

Mabogunje, A. (1981) 'Objectives and Rationales for Regional Population Redistribution in Developing Countries' in UN Population Distribution Policies in Development Planning. New York: UN Ch. II, pp. 19-29

McCall, M. (1983) Resource Planning Implications of Tanzania's Villagisation Programme. Enschede: Twente University of Technology

McCracken, K.W.J. (1981) 'Analysing Geographical Variations in Mortality: Age-Specific and Summary Measures', Area, 13, 3, pp. 203-210

McGlashan, N. (ed.) (1977) 'Studies in Australian Mortality', Environmental Studies Occasional Paper No.4 (Univ.of Tasmania)
McMaster, D.N. (1961) 'Change of Regional Balance in the Bukoba District of Tanganyika. Tanganyika Notes and Records, 56, 79-92
Mamdani, M. (1972) The Myth of Population Control: Family, Caste, and Class in an Indian Village. London: Monthly Review Press
Maro, P. & W.F.I. Mlay (1982) 'Population Redistribution in Tanzania' in Kosinski and Clarke, Ch. 24, pp. 176-81
Martin, L.G. & J. Trussell, Et al. (1983) 'Co-Variates of Child Mortality in the Philippines, Indonesia and Pakistan: An Analysis Based on Hazard Models', Population Studies, 37, 3 (November), pp. 417-433
Medawar, J. and Pyke, D. (eds.) (1971) Family Planning, Harmondsworth: Penguin, pp. 56-7
Moseley, M.J. (1979) Accessibility: The Rural Challenge, London: Methuen
Nag, P. (1984) Census Mapping Survey. New Delhi: Concept
O'Connor, A. (1983) The African City, London: Hutchinson University Library
Odingo, R.S. (1971) The Kenya Highlands: Land Use and Agricultural Development, Nairobi: East African Publishing House
Office of Health Economics (1972) Family Planning in Britain, Luton: White Crescent Press
Office of Population Censuses and Surveys (1978) 1977 Demographic Review: A Report on Population in Britain. Series DR No.1. London: HMSO, (prepared by Mr. E. Grebenik)
Office of Population Censuses and Surveys (1982) British Society for Population Studies Conference: Population Change and Regional Labour Markets, Occasional Paper No. 28. London: OPCS
Pacione, M. (ed.) (1983) Progress in Rural Geography. London: Croom Helm
Page, H.J. & Lesthaeghe, R. (eds.), Child-Spacing in Tropical Africa: Traditions and Change, London: Academic Press
Peach, G.C.K. (1968) West Indian Immigration to Britain: A Social Geography, London: Institute of Race Relations/ Oxford University Press
Peil, M. (1971) 'The Expulsion of West African Aliens', Journal of Modern African Studies, 9, 205-29
Phillips, D. & A. Joseph (1984) Accessibility and Utilisation: Geographical Perspectives on Health Care Delivery. London: Harper and Row
Preston, S.H. & Bennett, N.G. (1983) 'A Census-Based Method for Estimating Adult Mortality', Population Studies, 37, 1 (March), pp. 91-104

Preston, S.H. & Mari Bhat, P.N. (1984) 'New Evidence on Fertility and Mortality Trends in India', Population and Development Review, 10, 3 (Sept.), pp. 481-503
Prothero, R.M. (1974) 'Foreign Migrant Labour for South Africa? International Migration Review, 8, 383-94
Prothero, R.M. and Gould, W.T.S. (1984) 'Population Geography and Social Provision', in Clarke, Ch.13, pp. 111-126
Rapp, A., Berry, L. and Temple, P. (eds.), (1973) 'Studies of Soil Erosion and Sedimentation in Tanzania, Research Monograph No.1, Bureau of Resource Assessment and Land Use Planning, University of Dar es Salaam
Rees, P.H. & Wilson, A.G. (1977) Spatial Population Analysis, London: Edw. Arnold
Rempel, H. & Todaro, M.P. (1972) 'Rural-to-Urban Labour Migration in Kenya', in S.H. Ominde & C.N. Ejiogu (eds.), Population Growth and Economic Development in Africa, London: Heinemann, Ch.28, pp. 214-31
Rhind, D. (ed.) (1983) A Census User's Handbook. London: Methuen
Robinson, W.C. (ed.) (1975) Population and Development Planning. New York: The Population Council
Rogge, J. (1982) 'Refugee Migration and Resettlement', in Clarke & Kosinski, Ch.5, pp.39-43
Rondinelli, D.A. and K. Ruddle (1978) Urbanisation and Rural Development: A Spatial Policy for Equitable Growth. London: Praeger
Ross Report (1973) Report of the Population Panel. HMSO, 1973
Sabot, R.H. (1979) Economic Development and Urban Migration: Tanzania 1900-1971 London: Oxford University Press
Santos, M. (1979) The Shared Space: The Two Circuits of the Urban Economy in Underdeveloped Countries. London: Methuen
Sauvy, A. (1974) General Theory of Population London: Methuen
Shepherd, J. et al. (1974) Social Atlas of London. Oxford
Silitshena, R.M.K. (1982) 'The Regrouping Policy in the North-East District of Botswana,' in Clarke & Kosinski, Ch. 27, pp. 202-8
Sirageldin, I., Salkever, D., and Osborn, R.W. (eds.) (1983) Evaluating Population Programmes: International Experience with Cost-Effective Analysis and Cost-benefit Analysis. London: Croom Helm
Smith, D.M. (ed.) (1982) Living Under Apartheid, London: Allen and Unwin
Spengler, J.J. (1983) France Faces Depopulation. Durham, N.C.: Duke University Press

Stohr, W.B. and Taylor, D.R.F. (eds.) (1981) Development from Above or Below? The Dialectics of Regional Planning in Developing Countries, Chichester: Wiley
Sutton, K. & Lawless, R. (1978) 'Population Regrouping in Algeria: Traumatic Change and the Rural Settlement Pattern', Transactions of the Institute of British Geographers, 3, 331-50
Swindell, K. (1979) 'Labour Migration in Underdeveloped Countries: The Case of Sub-Saharan Africa', Progress in Human Geography, 3, 239-59
Tanzania, United Republic of (1981) Long Term Perspective Plan 1981-200. Dar es Salaam: Government Printer
Thomas, I. (1972) 'Infant Mortality in Tanzania', East African Geographical Review, 10 (April), pp. 5-26
Thomas, I.D. (1982) 'Villagization in Tanzania: Planning Potential and Practical Problems', in Kosinski and Clarke, Ch. 25, pp. 182-91
Thomas, I. (1983) Population and Health Facilities in Tanzania 1978. Vol. 1 Method and Results. Norwich: Overseas Development Group
Tietze, C. (1983) Induced Abortion: A World Review 1983. 5th edit. New York: The Population Council
Todaro, M. P. (1971) 'Income Expectations, Rural-Urban Migration and Employment in Africa', International Labour Review, 104, 387-413
Todaro, M. P. (1976) 'Urban Job Expansion, Induced Migration and Rising Unemployment', Journal of Development Economics, 3, 3
Todaro, M. P. (1976) Internal Migration in Developing Countries: A Review of Theory, Evidence, Methodology and Research Priorities. Geneva: International labour office
Trewartha, G.T. (1965) Japan: A Physical, Cultural and Regional Geography. London: Methuen
Twaddle, M. (ed.) (1975) Expulsion of a Minority: Essays on Ugandan Asians, London: Athlone Press
United Nations (1971) The World Population Situation in 1970. ST/SOA/Series A/49. Department of Economic and Social Affairs Population Studies No. 49. New York: UN
United Nations (1973) Determinants and Consequences of Population Growth. New York: UN
United Nations (1981) Population Distribution Policies in Development Planning, ST/ESA/SER.A/75, Dept. of International Economic and Social Affairs Population Studies No. 75. New York: UN
United Nations (1983) Indirect Techniques for Demographic Estimation. Manual X. New York: UN
Wander, H. (1969) 'Policies and Implementation Methods in the Internal Redistribution of Population: A Comparison of Different National Approaches to Guide the Flow of

Internal Migrants', Proceedings of the International Population Conference, London, 1969. Vol. 4, pp.3024-36
Warnes, A.M. (ed.) (1982) Geographical Perspectives on the Elderly. New York: Wiley
Warnes, A.M. & Law, C.M. (1984) 'The Elderly Population of Great Britain: Locational Trends and Policy Implications', Transactions of the Institute of British Geographers, 9, 1, pp.37-59
Warwick, D.P. (1982) Bitter Pills: Population Policies and their Implementation in Eight Developing Countries. Cambridge: Cambridge Univ. Press
Watson, W.B. & Lapham, R.J. (eds.) (1975) 'Family Planning Programmes: World Review 1974', Studies in Family Planning, 6, 8 (August), pp. 205-322
Webb, J.W. Naukkarinen, A. and Kosinski, L.A. (eds.) (1981) Policies of Population Redistribution. Oulu: Geographical Society of Northern Finland
White, P. and Woods, R. (eds.) (1980) The Geographical Impact of Migration. London: Longman
Wilson, M.G.A. (1979) 'Infant Death in Metropolitan Australia 1970-1973', Canadian Studies in Population, 6, pp. 127-44
Wolfson, M. (1978) Changing Approaches to Population Problems. Paris: OECD
Wolpert, J. (1965) 'Behavioural Aspects of the Decision to Migrate', Papers of the Regional Science Association, 15, pp. 159-69
Woods, R. (1982) Theoretical Population Geography, London: Longmans
World Bank (1974) Population Policies and Economic Development, A World Bank Staff Report. Baltimore: Johns Hopkins Univ. Press (for the World Bank)
World Bank (1984) World Development Report 1984. New York: Oxford University Press
Zachariah, K.C. & Kurup R.S. (1982) Determinants of Fertility Decline in Kerala, Paper delivered to the BSPS Conference on India's Population, Oxford, December 1982. (see Dyson and Crook, 1984)
Zelinsky, W. (1971) 'The Hypothesis of the Mobility Transition', Geographical Review, 61, pp. 219-49

Chapter Six

INTERNATIONAL MIGRATION: A SPATIAL THEORETICAL APPROACH

J. Salt

INTRODUCTION

The study of migration has become the cornerstone of the geographical interest in population, as a wealth of research testifies. Surprisingly, relatively little of that effort has been addressed to international movement, with certain major exceptions. Quite why this should be so is puzzling, but it can be suggested that it reflects the relatively scant attention to international migration paid by social scientists generally until the last couple of decades. It is only since about 1970, with large flows in Europe being particularly influential, that the subject has been widely regarded as a 'problem'. Even then, perception of it as mainly a labour market problem has ensured the dominance of political-economic approaches to causation.

Undoubtedly, a major constraint in the study of international migration is the availability of data, most of which are inaccurate, irregular and lacking in detail. The data problem is inextricably linked with the definition of migration. Every border crossing is not automatically designated a migration and both the selected time period for definition and the specified list of trip types considered as migrations vary from country to country. For researchers an ideal world would be one in which data are available on all border crossings so that all possible types of movement are recorded. In reality governments collect data only on certain types of immigrants, usually those to whom they attribute some element of permanency and/or where there is some intention to live and work in the host country. The result is that the range of data collected by states varies enormously and studies of international migration have tended to reflect the definitions used. In one sense, then, administrative decisions about data collection have imposed their own form on the researches carried out.

In the past, when fertility was high and flows of international migrants were relatively small elements in population change, data problems had less demographic importance than they do today. Now fertility is almost universally falling, while international movements are frequently large in a fixed matrix of independent states. The behaviour of international population flows can be highly influential in a country's demographic development. Already in some Middle Eastern countries the indigenous population is in a minority. Even where the scale of flow is small the quality, especially the degree of skill, may be very important. All states wish to gain high level manpower, while these days a decreasing number want unskilled workers; even fewer now want to attract permanent settlers.

The list of potential types of border crossing that could be construed as international migration highlights the nature of the definitional dilemma, and its associated data-gathering exercise, even for countries with good statistical services. Migrants may be permanent settlers; long or short term contract workers, seasonal or casual workers; they may be legal or illegal; increasingly attention is being paid to 'transients', especially specialist workers transferred for short periods (including diplomats, those working for multi-national organisations, and those engaged on specific projects) but also including tourists, businessmen and students. In addition to these voluntary migrants are those who have less say in the move, including members of the Armed Forces and refugees - the latter group falling into several categories, as will be seen later.

The reality today is that growing interdependence between a large number of states operating in a global economy has led to a widespread international population mobility within continuums of time and space. This cannot easily be accommodated by traditional classifications of international migration, based on arbitrary breakpoints, which result in some types of movement which are regarded as worthy of study, usually because data on them are collected, and others which are not.

Thinking about international migration needs, therefore, to be in the context of a flexible system in which a range of actors - migrants, their families, those left behind, origin and destination governments and populations - are active participants. The role each plays depends not only on interaction with other elements in the system, but on the outside influences which affect each element individually. Such a system will consequently have some characteristics in common with other systems, but will respond also to particular local events (Mabogunje, 1970).

The empirical nature of much of the work by geographers on international migration makes a review of that work tempting. This chapter will, however, be more concerned with matters theoretical, and particularly with an evaluation of spatial approaches to the determination of international movement. It will suggest an analytical framework of growing importance, based on twin pillars. The first is that geographical patterns of international migration occur within a systems framework and are related to complex networks of interaction. The second is that combinations of global and local forces produce spatial migration networks that differ from each other in scale, nature and behaviour. All this may seem elementary to the geographer. However, a hard look at theories of causation developed by other disciplines, especially political economics, suggests a tendency to assume their universal applicability. In the absence of empirical testing in a range of geographical circumstances such assumptions can lead to misleading conclusions about how migration patterns have evolved, are currently functioning, and will develop. Much of the chapter will focus on labour migrations, but the analysis will be extended to include refugees as well.

A Spatial Approach to Migration Theory

The study of international migration is replete with theory, and scholars from a range of disciplines have sought general laws governing the causes and consequences of movement. One result of this search for general order has been a diminishing awareness of the geographical variability of international migration.

Insofar as there is a spatial theoretical approach to international migration, it is based on the concept of diversity. Economic development is uneven over time and space because of the irregular distribution of raw materials and energy, agglomerations and economies of scale inducing core-periphery dichotomies, uneven diffusion of technology and distribution of political and economic power, and the cyclical nature of modern economies. Migration is itself a variable response to this spatial diversity in the means of production, not surprisingly taking on a range of forms manifest in causation, character and impact.

Geographers have traditionally regarded migration as a spatial re-allocation of human resources. Their approach has emphasised the diversity of the process and its effects upon origin and destination areas. Migration is ultimately studied as one element in general regional change, the approach both empirical and pragmatic, relying on whatever disciplinary theoretical frameworks are appropriate for the problem under consideration.

At the core of spatial theory is, of course, distance. In much migration theory the distance concept is central and macro-analytical models derived from social physics have provided useful descriptions of complex flow patterns and frequently indicated fruitful lines for further analysis. Gravity and regression models, for example, have shown how average distance travelled by migrants can be related to levels of development - the distance coefficient in the gravity model normally declining as economic development proceeds. Regression models can indicate the strength and aggregate relationships between migration and other variables, but their greatest value is often the insights that come from analysis of the residuals from the models and which can point to other, perhaps unsuspected, influences. One weakness of models based on distance is the rather nebulous nature of the concept itself, for it can be measured in different ways: in terms of simple mileage, cost, time or intervening opportunities. Increasingly it is realised that perceived distance may well be a more valuable concept than linear distance.

Over the last twenty years particularly, there has been a move away from models of migration derived from social physics where migration is regarded as the outcome of impersonal macroscopic laws, to a more microanalytical approach which seeks to comprehend the processes involved in creating the spatial pattern. Individual migration behaviour is now regarded in much geographical work as an expression of a decision-making process that need not be economically or spatially rational. Concepts like distance decay and intervening opportunities have become secondary to the identification of specific causes of population and regional change and to the interrelationships between social processes and individual values, perceptions and preferences (Wolpert, 1965).

In both these approaches a fundamental problem is the geographical scale chosen for analysis. In fact, for many migration studies space cannot be regarded as a continuum, movement being recorded between discrete spatial units at a variety of scales from local to international. Where a distinction is made between internal and international movement it reflects the compartmentalisation of the world into nation states which in turn determines the recording of movement. Hence there can be relatively short distance movements which are international and much longer ones that do not cross national borders. In this strict sense distance is often irrelevant when considering international movement. Indeed, most of the major features of migration are common to both internal and international movement. Spatial inequalities can be held to account for movement over any distance; between areas broadly equal in levels of living, migration of highly skilled people may or may not be across national borders. There is evidence of distance decay at both internal and international scales: in the latter case, for example, the tendency for Italians and Yugoslavians to concentrate in southern Germany, Algerians in Mediterranean France. Selectivity principles operate, streams may be self-supporting and reverse flows occur at all scales. This means that the main distinguishing characteristic of international migration is the political dimension, with relationships between pairs of places strongly influenced by political ideologies and with governments frequently taking an active role in organising movement. Political barriers become additional hurdles for anyone moving across the space continuum, and the international migrant is someone who has the power to cross them.

The Nature of Political Barriers

The nature of these barriers lies in the sovereignty of the state, which results in laws and regulations governing citizenship and the rights of entry and exit, and which reflects the development of attitudes towards foreign migration (United Nations, 1982). As far as international mobility is concerned regulations may take the form of laws, administrative measures and agreements between states. Agreements may be reciprocal; they may also be bilateral or multi-lateral. Labour flows into northwest Europe in the 1960s and early 1970s were governed by a network of bilateral treaties which had the effect of creating a vast labour market extending north to Finland and south into tropical West Africa, and encompassing most of the Mediterranean basin. Northwest Europe has two sets of multi-lateral agreements, so far not successfully replicated elsewhere in the world: the Nordic and EEC common labour markets. Within each of these, member nationals have the right freely to cross frontiers to seek and take up work. In practice mobility has never been entirely free, as each state has retained powers to refuse entry into selected employment to non-nationals.

Where entry is not governed by supra-national agreement, a variety of control options is available to potential host governments. Mostly these operate through the issue of residence and work permits and may be either short or long term. In addition a medical certificate may also be required, a stiff hurdle for some potential migrants but a way of using the employment insurance system to advantage by the host country - healthy immigrants pay insurance contributions but are less likely to require medical services. The permit system allows considerable flexibility and is increasingly used in some countries selectively: family members often find it more difficult than breadwinners to enter, while more highly skilled immigrants can usually gain entry more easily as countries seek to enhance their stocks of human capital. Today brains are more in demand than brawn, one result of which is that some origin countries are now actively persuading their immigrant citizens with needed skills to return. Entry barriers present particular problems in some areas, as in parts of Africa, for example, where colonial frontiers cut across homogeneous ethnic groups, though they may be irrelevant when refugee movements occur. Quite frequently the very existence of entry controls gives rise to illegal migration, a growing problem in much of the world where tighter entry criteria actually encourage proportionally more clandestine movement amongst those who feel it is pointless trying to gain legal entry because they lack necessary qualifications.

Government attitudes towards regulations on departure may also create barriers, although increasingly today some states are adopting policies designed to make it easier for their nationals to leave. Article 13 of the Universal Declaration of Human Rights stipulates that everyone has the right to leave any country, including his own. In reality some countries, especially in the Communist block, do prohibit emigration, either generally or for certain categories of citizen, control being exercised through the issue of necessary documents, including passport, visa of destination country, and vaccination certificate. More common now are policies aimed at stimulating emigration either for relieving demographic pressure, as in the case of Rwanda, or for the economic and social benefits stemming from remittances and the reduction of unemployment. A growing tendency, especially in southern Asia, is the organisation of emigration through contracts. Countries may allow potential host employers to come in and recruit, as in the case of China, use their own Ministries of Labour to do the recruiting, as in Thailand, Sri Lanka and Bangladesh, or set up their own bodies to organise recruitment and emigration, for example, the Philippine government's Employment Development Board which recruits workers whose skills are needed for contracts signed between itself and other governments (Abella, 1984).

NETWORKS

International Migration Networks

In seeking to understand how international migration occurs and evolves it is important to review the principal networks that exist. What soon becomes clear is their diversity of geographical form and characteristics and, therefore, that attempts at explanation must regard them as systems with varying attributes and organisational forces.

The migration network of Western Europe occurs in a region of considerable distinctiveness. The countries there are industrially advanced with highly interacting economies and with active frontier regions. National migration policies vary, but have generally evolved within democratic systems of government which have sought a consensus on the issue. To the south a gradient occurs in standards of living which is traversed by large numbers of workers and their families. Western Europe is unique in that ex-colonial links have led to distinctive patterns of immigration in several cases, with flows being dominated by the residue of empire: the inflow of *pieds noirs* into France, for example, fits uneasily into any model of migration. Schemes for free movement of population have also been developed between associations of states.

Within the Western European network marked differences occur in rates of immigration and in the directions of links between origin and destination countries. France and West Germany both have similar numbers and proportions of immigrants but draw from different sources. In Great Britain, levels of immigration have remained remarkably low and sources have generally differed from those of continental countries. Patterns have changed over time too: for example, in the later 1950s the Italian flow switched from France to West Germany and Portuguese and Spaniards were sucked into France instead. In recent years rates of inflow and outflow have varied by national group: in West Germany after 1975 inflows of Turks and Italians rose sharply in comparison with other groups and though falls in outflow were common to all nationalities, again there were some divergences of experience (Salt, 1985).

The most active migration network at present is that centred on the Middle East, but it differs considerably from that based upon Western Europe and cannot be regarded as merely a diversion of flows from former European sources (Birks and Sinclair, 1980; Seccombe and Lawless, 1985). The destination countries are tiny in population terms. Hence the influx of large numbers of foreign workers has led to labour forces dominated quantitatively and qualitatively by aliens (Stahl, 1982). The Middle East countries are still largely non-industrialised, their economies based on one commodity - oil - which makes them potentially highly vulnerable. Cultural influences are non-European and there is a lack of tolerance in religious attitudes. Political orientations span a wide spectrum in the area as a whole; although most labour importing is done by countries with a broadly pro-western capitalist stance, forms of government are not democratic. In general, host countries have looked eastwards for their labour, latterly over very long distances, to Thailand and the Philippines, for example. Although there have been attempts to recruit Moslems from Jordan, Turkey and North Africa, many workers in the 1970s expressed other faiths, creating a religious and ethnic divide between the foreign and the indigenous populations. Lack of industrialisation has meant that foreign workers were employed especially in low-skill jobs in construction rather than in the manufacturing and service industries which many of them had filled in western Europe. Gulf states too have set their faces against settlement, labour recruitment being managed strictly on a rotation basis, no doubt having learned from Europe. As a result flows have become increasingly organised, with large numbers of skilled and contracted expatriates, and the bulk of labour being recruited and managed by agencies or through government recruitment offices. Origin countries too tend to be at lower levels of economic development and more diverse than those of southern Europe. A growing distinction is emerging between newly industrialised countries (NIC) like South Korea and the Philippines, where the growth of large multi-national enterprises is promoting a corporate package style of operation, and others like India, Pakistan and Bangladesh where migration, declining in amount, is much less well organised and is not part of wider business relations (Abella, 1984; Ling, 1984).

The labour migration network centred upon the USA and Canada differs from those described above. Levels of living gradients to the south are very steep, giving rise to a continuing problem of illegal movement (Keely and Elwell, 1981). Furthermore, the immigration tends to have a marked ethnic character leading to major integration problems (e.g. Portes, 1981). Political relationships are especially important affecting flows between the USA and a number of Caribbean and central American states and even in some cases generating flows. Since the USA and Canada are in little need of further supplies of unskilled labour, except in certain circumstances such as the seasonal immigration of agricultural labour in the south-west (Ranney and Kossoudji, 1983), the emphasis is on the import of skilled workers but these are often those most needed by origin countries themselves. Morales (1983) has pointed out the transitional role of less-skilled undocumented aliens in increasing the flexibility of employers in highly developed industries such as car manufacturing, who seek to reduce employee numbers and substitute capital for labour.

Within the Caribbean there may be a step-like quality in labour migration with end points in the USA, Canada, Venezuela, Argentina and Brazil, but over the last couple of decades there has undoubtedly been a general enlargement of the region's migratory system which has turned several countries into sources of multiple outflows (Bach, 1983; Kritz, 1981). Colombia exports migrants to Venezuela, Ecuador, Panama and the USA; Salvadorians have been going to Honduras for decades and now go to Nicaragua, Guatemala, Costa Rica and Panama. Guatemalans go to Mexico, Haitians to the Dominican Republic and Bahamanians and Windward Islanders to Trinidad and Tobago. Venezuela receives migrants from Colombia, Argentina, Peru and the Dominican Republic; Mexico imports Guatemalans for its tobacco and coffee harvests and has been receiving refugees from throughout Central America. Puerto Rico, for long the source of immigration to the USA, has become a recipient for significant numbers of migrants, mostly from the Dominican Republic and Cuba. Furthermore, these flows are variable in composition. Most prevalent are people of low skill, mostly male, who circulate temporarily across continuous borders, but growing numbers constitute the brain-drain/gain to the USA or are refugees (either rural peasants displaced in Central

American wars or better-educated professionals fleeing repressive regimes such as those of Cuba, Chile or Argentina). The USA - Canada - Caribbean region migration network is thus highly complex geographically, and its pattern will only be understood in terms of the interaction of both general forces and highly variable local conditions.

Beyond these major networks others have their own particular characteristics. Marshall (1981) has explained the preferential orientation to Argentina by neighbouring states in terms of a range of geographical, historical and cultural links. These include the socio-cultural homogeneity with Uruguay; the existence of established communities of each major source nationality in Argentina already, so that immigrants are not cultural newcomers and join reception systems developed already; the historical dependence of Paraguay upon Argentina, so that labour flows merely reinforce what already exists in economic links generally; the easy communication across the fluid frontier regions with Bolivia, Paraguay and Chile; and the common origins and traditions linking Paraguaians and north-eastern Argentinians, Bolivians and north-western Argentinians.

The migration network centred upon South Africa has been institutionalised over a long period and is clearly distinctive from those elsewhere (Böhning, 1981). It is a highly organised system, designed specifically for rotation rather than settlement, and maintained by an individual political philosophy that governs the behaviour of the network. Political relationships with source countries have determined marked changes in flow patterns in recent years, for example, the virtual cessation of movement from former Portuguese territories, Botswana, Zimbabwe and Zambia, and the increased reliance on 'internal' sources.

A very different type of network exists within Eastern Europe where labour imbalances are less easy to explain in terms of capitalist forces and the need for the international migration of labour is not supposed to exist (Guha, 1978). Numbers moving seem relatively low, though in the absence of suitable data one cannot be sure (Schultz, 1975). Technical co-operation is the principal reason for the movement with most migrants being specialists of one sort of another, often in construction. East Germany, Czechoslovakia, Hungary and the USSR are the principal host nations, Poland and Bulgaria the main sources. Yugoslavia, of course, constitutes a major anomaly - its links have been mainly with Western Europe since the 1960s - but it seems also that other eastern Europeans, Poles and Rumanians especially, worked in the West in the 1970s: one estimate for 1972 was 4,000 (Schultz, 1975).

A final network to consider is that centred on Australasia. Here an over-riding consideration is geographical remoteness making it impossible to tap rotational sources easily. Immigration has been mainly European, and for purposes of settlement rather than for labour market reasons (Zubrzycki, 1981). Political attitudes have also militated against more local supplies from Polynesia or south Asia.

The Links between Origin and Destination

Such are the complexities of the networks outlined above, that we can only begin to comprehend them by looking at parts of them individually, and gradually see how each interacts with the rest to create the whole network.

Most attempts to explain geographical patterns of international migration have been at the macro-scale, invoking economic ideas of push and pull formulated into regression models (Salt, 1981). The models have normally sought to account for temporal variations in migration rate between pairs of origin and destination countries. Such analyses have demonstrated the significance, at the aggregate level, of labour market conditions (e.g. Böhning, 1970; Drettakis, 1975; Lianos, 1975; Wadensjö, 1978). Unfortunately, lack of detailed flow data and the disciplinary interests of researchers, mostly non-geographers, have hindered though not precluded attempts to model the complete matrix of movement for any migration network.

Such macro models tell us little about the geographical reality of much international migration, including the varying propensities to migrate to and from particular places, the links between local origins and destinations created by flows of information and the ways in which these change over time. Migration rates and composition vary considerably by region in both origin and destination country and to link and explain them requires consideration of many variables within some kind of systems framework. Only by so doing can the interaction inherent in the migration process be fully understood.

A number of studies have sought to explain regional variations in immigration rates in terms of a range of socio-economic and demographic variables (e.g. Baucic, 1977; Jones, 1964; King and Strachan, 1979, 1980; McDonald, 1969; Parenti, 1958). Other studies have demonstrated the ways in which links have developed over time between the specific regions; for example Jones (1973) and King (1977) for Malta, Tapinos (1966) for Galicia, Augarde and Prevost (1970) and Simon and Noin (1972) for Algeria. In so far as some of these studies seeking to explain emigration rates refer only to specific time periods their use is limited by the static framework used. They do, however, indicate the main forces involved and their respective strengths. Perhaps more important, analysis of the geographical pattern of residuals can give important clues why some places err from the norm. King and Strachan (1980), for example, were able to demonstrate a clear relationship between the pattern of residuals to their model and specific elements in the local economy.

In recent years there has been a growing interest in return migration. Much of the discussion has focused on developmental issues such as the use of remittances and savings and the value of skills acquired (or lost) while abroad. The theoretical approach most adopted has been economic, regarding return migration as an agent of the development process. It has assumed a degree of homogeneity among return migrants that is not replicated when local areas of return are considered. Furthermore, the different motivations of those coming back, compared with those going out, are not clearly distinguished, largely it must be admitted, because there is still little empirical information. Yet the evidence does indicate that motivations do vary. The study of King, Strachan and Mortimer (1985) in southern Italy demonstrated the importance of non-economic considerations for returnees, a significant finding in view of return migration's hypothesised contribution to economic change in regions of under-development. It implies, for example, that they are unlikely to be too concerned about investment that will be good for their home region's economy.

Complexity is also evident when the detailed geographical pattern of return migration is examined. Return migrants to north eastern Italy were much more likely to continue working in the sector they had occupied while abroad, and more easily able to obtain jobs on return, than those returning to the south, the differences reflecting regional variations in the patterns of industrialisation and development (Saraceno, 1985; King, Strachan and Mortimer, 1985). In Portugal, Lewis and Williams (1985) discovered significant differences in geographical patterns of return settlement between emigrantes (returning from the industrial countries to the north) and retornados (returning from Portuguese African territories but born in Portugal) and also differences in the characteristics of both groups between those returning to urban areas and those to rural areas. For example, emigrantes were more likely than retornados to return to rural places of birth so their pattern of urban drift was less pronounced; richer retornados were more likely to return to urban areas, to have their own businesses and employ more people than poor retornados and emigrantes. Emigrantes returning to more rural areas spent less on housing and more on land and were more likely to be pushed back to work than emigrantes returning to urban areas; for example, 47 percent of rural area emigrantes intended not working after return but only 25 percent managed it, compared with 25 percent and 26 percent respectively of the large city returnees.

Even at the level of individual settlements there is a distinct geographical pattern of return. A number of studies have shown how certain parts of settlements have come to be dominated by the houses of returnees, the detailed patterns resulting from a range of social and economic factors. Findlay and Samha (1985) have demonstrated a similar effect in Amman during the 1970s.

High Level Manpower Migration

The inadequacy of gravity and regression models, and the need to view migration in systems terms, is evident in studies of the movement of highly skilled people. Explaining their migration presents some new problems, for spatial disparity is relatively unimportant to people moving for career reasons in an increasingly global economy.

Such moves occur not only within the migration networks centred upon the older post-industrial economies in Western Europe and North America, but are increasingly found in newly industrialised countries and in those still going through the early phases of the industrialisation process (Lawless and Seccombe, 1984; Ling, 1984; Royal Scientific Society, 1983; Stahl, 1984). Using data from the International Passenger Survey, Findlay and Godden (1985) have demonstrated a significant change in the character of British migration to the Middle East. In the early-middle 1970s manual and clerical workers were heavily in demand, as infrastructure projects started up in the Gulf. From the late 1970s fewer of these workers have been recruited (partly owing to competition from Far East corporations and labour sources) but there has been an increased demand for professional and managerial staff, especially those with scientific and engineering skills and, increasingly, those in education and the welfare sector. In 1982 63 percent (13,316) of all workers recruited from Britain to the Middle East were classed as professional or managerial.

There is as yet little empirical evidence on the nature of this movement, but an examination of recent data on the issue of long-term work permits in Britain provides some interesting clues. Ideally Table 6.1 should include EEC countries but the free movement of labour provisions of the Treaty of Rome mean that permits are not needed, and there is no other data source. It is clear from the figures for the major immigrant sources that although there is some variation between countries, the most permits issued in each case are for professional, managerial and related staff, and that during 1981-83 the proportions were fairly constant.

This group does not constitute an amorphous and undifferentiated mass. Table 6.2 shows, for each country in 1983, the breakdown of the professional, managerial and related immigrant group into its constituent parts. There are clear variations between countries in the proportions of different types of these high level workers, and some origins are much more important than others for specific skill types. Japan and Switzerland, for example, send a high proportion of pro-

Table 6.1: United Kingdom: Long-term Work Permits Issued 1981-83

Source	1981 Total Nos.	1981 Prof. Managerial related	1982 Total Nos.	1982 Prof. Managerial related	1983 Total Nos.	1983 Prof. Managerial related
Austria	44	91	33	91	35	86
Japan	731	81	764	84	806	85
Norway	128	91	129	95	94	91
Finland	51	92	41	85	79	90
Sweden	159	88	129	89	226	93
Switzerland	74	85	91	87	60	88
U S A	1958	94	1945	94	1862	94
Australia	264	93	203	95	267	92
Canada	187	96	152	94	235	92
Greece	57	91	46	96	52	94
Turkey	47	89	56	82	36	81
India	168	84	134	86	162	84
Malaysia	203	98	250	98	257	98
Total for all countries	5780	88	5569	88	5818	88

Source: Department of Employment, Overseas Labour Division (Unpublished).

fessional and managerial support staff to their general (senior) management, but few health specialists. For Norway, scientific, engineering and technological occupations are the most important, although this group also tends to be significant from the less developed countries, including Turkey, India and Greece. Migration from Malaysia is dominated by health professionals. As a general rule, the more highly economically developed the origin country, the higher is its proportion of general (senior) management and support staff, though there are exceptions. Quite why these variations occur is not immediately clear, but they almost certainly reflect such variables as government policies (e.g. in using overseas medical staff in the Health Service), the nature of economic links and the recruitment and staff transfer policies of individual employers.

Table 6.2: United Kingdom: Long-term Work Permits to Professional and Managerial Workers 1983. (Type of professional/managerial staff (key below)

	1	2	3	4	5
Austria	26.7	33.3	13.3	16.7	10.0
Japan	20.4	37.9	1.5	17.1	23.1
Norway	12.9	25.9	2.3	47.1	11.8
Finland	16.9	36.6	11.3	16.9	18.3
Sweden	27.5	44.5	0.5	18.0	9.5
Switzerland	11.4	54.7	1.9	11.3	17.0
U S A	19.4	40.1	8.7	24.5	7.3
Australia	11.7	32.8	27.5	21.9	6.1
Canada	12.9	35.9	8.8	32.3	10.1
Greece	2.0	30.6	6.1	46.9	14.3
Turkey	3.4	24.1	6.9	51.7	13.8
India	0.7	30.1	11.0	37.5	14.0
Malaysia	0.0	6.3	86.2	7.5	0.0
Total for all immigrant source countries	14.5	33.6	18.7	22.9	10.3

Source: Department of Employment: Overseas Labour Division (unpublished)

Key:
1. General (senior) management.
2. Professional, managerial and related support staff.
3. Professional/managerial: health and welfare.
4. Professional/managerial scientific, engineering and technological staff.
5. Other managerial.

The System and the Career
What has happened is that the interplay of highly skilled labour forces with advanced industrial economies in an interacting urban system has generated 'brain exchanges'. Migration theories based on concepts of distance, ideas of push and pull, and gradients in wages and standards of living, are inappropriate for explaining these exchanges.

An explanatory framework for high level manpower migration should be based on the disaggregated nature of the modern labour market, in which specialist skills and training mean that the workforce is segmented into self-contained non-competing groups (Salt, 1983). Attempts at explanation must distinguish specific occupational types and examine their relationship with labour market processes and institutions. A close association exists between the career path of the individual, the nature of the job and the migration demands imposed by the organisation of work and the internal structure of the employer.

Within the migration system that results the concept of career is most important. A career consists of a sequence of jobs held by an individual and related to each other by the acquisition of skill and experience. Mobility between jobs results from either task or locational change and may occur within an employing organisation or in movement between organisations. The career path can then be defined as the route taken by the employee through the sequence of jobs (tasks), occupations (collections of tasks), employers and locations. The choice of route broadly conforms to the idea that a career will progress upwards. We may hypothesise that on these career paths critical points will occur at which propensity to move increases and labour migration results. These are points which primarily reflect the nature of the occupation and the structure of tasks it contains, and the way in which an employer organises work and manages careers. The length and nature of career paths vary, and the interconnections reflect the organisation of work by the employer: for example, whether the corporate philosophy is that careers should be functional or general.

From the demand side the employer has a large system of fixed jobs into which must be fitted eligible people. Different policies exist for doing that. But such interaction cannot exist in isolation from supply side characteristics in the system, particularly attitudes to work and employee behaviour. There is some evidence that these attitudes vary between migration systems: American managers seem far more mobile than their European counterparts, and Jennings (1971), has written of them as 'mobilcentric'. Hence, as an individual moves through his career his decisions about where to work, and what at, are affected by constraints and influences which operate at particular moments in time.

Geographical migration patterns are, therefore, determined on the one hand by the location decisions of employing organisations and the spatial division of labour they favour, and on the other by a group of eligible people with degrees of skill and experience already acquired. These are elements in a system the energy for which is provided by the need for employers to fill vacancies with the right sort of skills, and by the desire of employees for careers which present possibilities for promotion, job satisfaction and general improvement in lifestyle.

Much of this movement takes place within the internal labour markets (ILM) of multi-locational organisations. Already in the UK it is estimated that over half of inter-regional labour migration involves people not changing employer (Gleave, 1983). For multi-national organisation ILMs have become international, with employees freely moving to jobs in foreign parts, for a range of reasons and in a variety of forms of corporate organisational structure. For ILM migration the organisation itself becomes the context in which the migration system develops. The geographical structure of the organisation, career development procedures and location of corporate functions (like R & D centres, head and regional offices) combine to fashion the geography of migration. The system is usually lubricated by generous compensation packages, including the use of specialist relocation services, which reduce the friction of both linear and psychological distance (Salt, 1983).

REFUGEES

The Nature of the Problem

Most research on international migration, and most of this chapter so far, has assumed that movement is largely voluntary. However, there can be no doubt that refugees now constitute a major element in international migration. Their numbers have grown from an estimated one and a half million in the aftermath of World War II to about ten million today (Crisp and Nettleton, 1984). Over the last 50 years both the scale and spatial focus of refugee movement have changed. In the inter-war period Europe was the centre of the problem; from the 1960s onwards new waves of refugees have emanated from Third World countries where instability, under-development and revolution have led to internal and external conflict. Most of them are to be found in Africa, the Middle East and southern Asia (Figure 6.1). In these new circumstances old remedies like repatriation, local settlement and resettlement have been revealed as grossly inadequate. In particular the refugee problem has become more politicised with cynical interest often vested in preserving the identity of ex-patriate

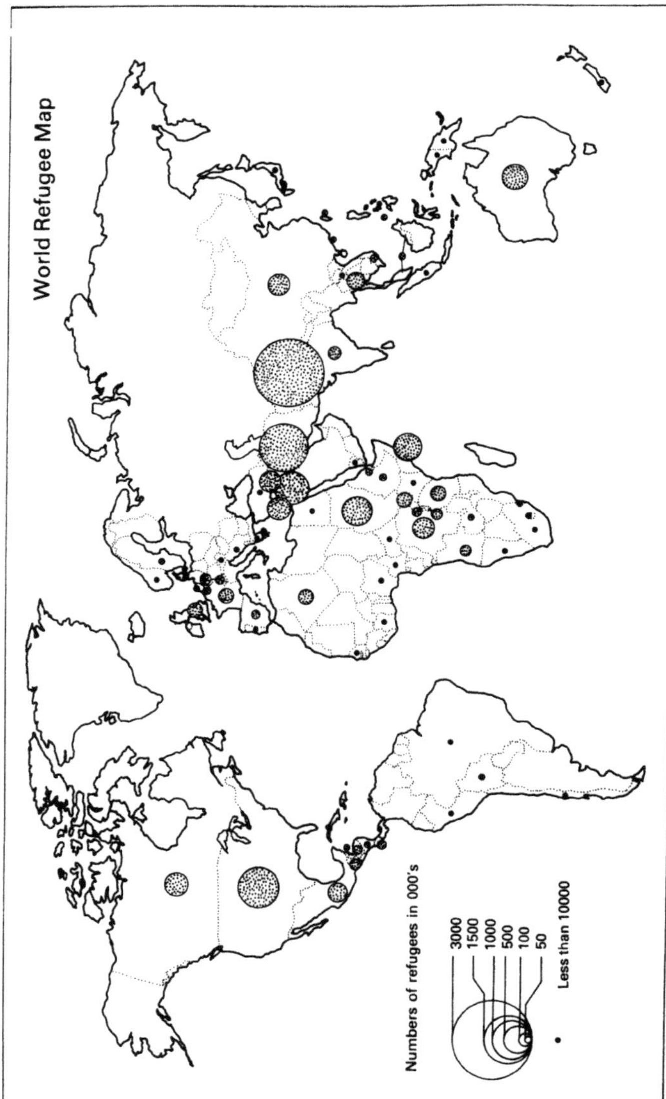

Fig. 6.1 Distribution of refugees 1984

groups, like the Palestinians in the Middle East and Afghans in Pakistan, allowing them to become active agents in fomenting and maintaining armed conflict with a view to further political change.

The association of refugees with conflict is reflected in the definitions used. The United Nations Convention of 1951 and the Protocol of 1967 Relating to Refugee Status define a refugee as someone who leaves or remains outside their own country owing to a well-founded fear of being persecuted for reason of race, religion, nationality, membership of a particular social group or political opinion. This definition has been widened by the Organisation of African Unity to include any individual who, owing to external aggression, occupation or foreign domination, or events seriously disturbing public order, is compelled to leave his habitual place of residence. This broader definition allows us to regard refugees as people whose presence abroad can be attributed to fear of violence.

Not surprisingly the scope of agencies concerned with refugees has also widened to include not only the refugees themselves, the origin countries and immediate destinations, but also more remote destinations, agencies striving to provide assistance, and major powers involved in the conflict directly or indirectly. Repatriation has become less a feature, local settlement less popular especially if the destination countries are already poor, and easily swamped by large numbers. Local settlement programmes are too often associated with agricultural settlement when the tide of population movement is towards urbanisation. Many states, while willing to accept refugees in the short term are unwilling to allow long term resettlement unless the numbers are small and special programmes can be derived for them, for example, that of the UK towards the Ugandans and Vietnamese boat people. Though often resettlement else where may remain a long term aim for refugees richer states have proved unwilling to accept them, especially in the recent recession. In any case the permanent resettlement of most African refugees outside Africa, for example, is just not feasible.

Refugees and Migration: A Theoretical Perspective

In the study of international movement the analysis of refugee flows, and particularly their determination has been divorced from concern with other forms of migration. To a considerable extent this reflects disciplinary interest. Migration specialists, mainly geographers, economists, sociologists, demographers and political scientists, have concerned themselves with labour flows and settlement, and have steered clear of moves that seem unpredictable and are triggered by political events. Specialists in international law, and some political scientists, in their studies of refugee flows have focused more on consequences than determination. Causative theories of refugee movement are, therefore, lacking.

It is only recently that very much attention has been paid to the need to produce a theory of refugee movement, though there have been attempts, usually rooted in sociology, at producing theories of refugee integration and settlement. It is, of course, possible to see elements of a refugee theory in some of the classic migration theories. For example, refugee movements are multi-step (Ravenstein, 1885), and refugees do have to surmount intervening obstacles (Lee, 1966). Petersen's (1958) general typology of migration recognises two basic refugee types, forced and impelled, though this grossly oversimplifies the process since refugee flows are immensely varied in both scale (geographical and volume) and time (short and sharp, long drawn out). With little theoretical basis having been developed, much of the writing on refugees has been pragmatic and descriptive, focussing either on flight or settlement and with the aim of informing policy makers, or framing advice for operational personnel dealing with refugee administration (Steins and Tomasi, 1981). One of the first attempts to provide a comprehensive conceptualisation of the determinants of refugee flows was that of Kunz (1973). He developed a model based on the distinction between "anticipatory" and "acute" refugee movements. The anticipatory refugee prepared for the move, has a clear knowledge of destination and permission to stay; he is not easy to distinguish from a voluntary migrant except that the push is usually more important than the pull. This type of movement Kunz refers to as push-permit. For the acute refugee the push motive is overwhelming, there is no certainty as to destination and the refugee may at first assume that the move is temporary. Kunz hypothesised three types of acute refugee sub-models, push-pressure-plunge (moving from the country of immediate asylum to a third country for permanent settlement), push-pressure-stay (in the country of asylum), push-pressure-return.

Associated with these patterns of flight and arrival is the introduction of stages in the process. The term "vintages" is used to describe the flight of successive groups as the political situation deteriorates. Vintages may unite to form "waves", creating cycles of refugee movement. Vintages are essentially departure and transit cohorts, displaced by flight, force or absence (defection). The Kunz model therefore relates associated departure cohorts (vintages) with the form of displacement, leading to an asylum interval (described as "midway to nowhere") and finally resettlement cohorts (waves). It presents us with a conceptual typology, a necessary first step in creating a theory of refugee migration. Kunz later (1981) elaborated his ideas by attempting to encompass the attitudes of refugees to events in their home country before movement, based on the degree of alienation experienced and, as far as resettlement was concerned, on the attitudes of the new host country towards such factors as cultural compatibility, tolerance, and immigration and asylum laws.

The Kunz model is an attempt to formalise the process of refugee movement through the development of a typology involving displacement, movement, settlement and the attitudes of refugees and asylum countries. It provides some conceptual clarity and is empirically based, thus allowing predictions to be made of the likely course of events once a movement is triggered. However, it lacks mechanisms and so has little explanatory power; it is particularly weak on the forces in origin countries that propel movement. It is only a first step in theory development, albeit a valuable one.

Growing interest among migration scholars in refugee problems during the 1980s is now leading to a serious attempt to develop conceptual models of refugee flow determination that move on from Kunz's typological attempt which focuses on refugees themselves, towards a linking of refugee movement with the political instability in which the origins of most refugee flows are rooted. The need for such a view has been expressed by Barber (1984) who has drawn attention to the role of super power rivalry in fomenting instability. Of particular interest and importance is the theoretical framework recently articulated by Zolberg (1985), based on the tensions and conflicts inherent in political transformation, and worth rehearsing in some detail. He argues that although the events that trigger refugee flows are unpredictable, they do not occur randomly but are manifestations of processes which are themselves related to structural features of contemporary world politics. The determinants of refugee flows are, therefore, amenable to theoretical analysis.

The basis of Zolberg's argument is that refugee flows tend to be generated by tensions and conflicts inherent in two major types of political transformation, an abrupt change of regime and/or the re-organisation of political communities. These conditions are exacerbated by the socio-economic environment which contributes to poverty and instability in many less developed countries, and to the susceptibility of many of them to the intervention of international political and economic forces. Such intervention, direct or indirect, tends to make conflicts more violent and to increase the probability that they would trigger refugee flows. It also means that an analysis of the determinants of refugee movement should be cast in a trans-national framework.

A Refugee System

Figure 6.2 shows the principal element in a global system of refugee flow determination. It contains three main elements: the origin countries of most refugee flows, mainly Third World

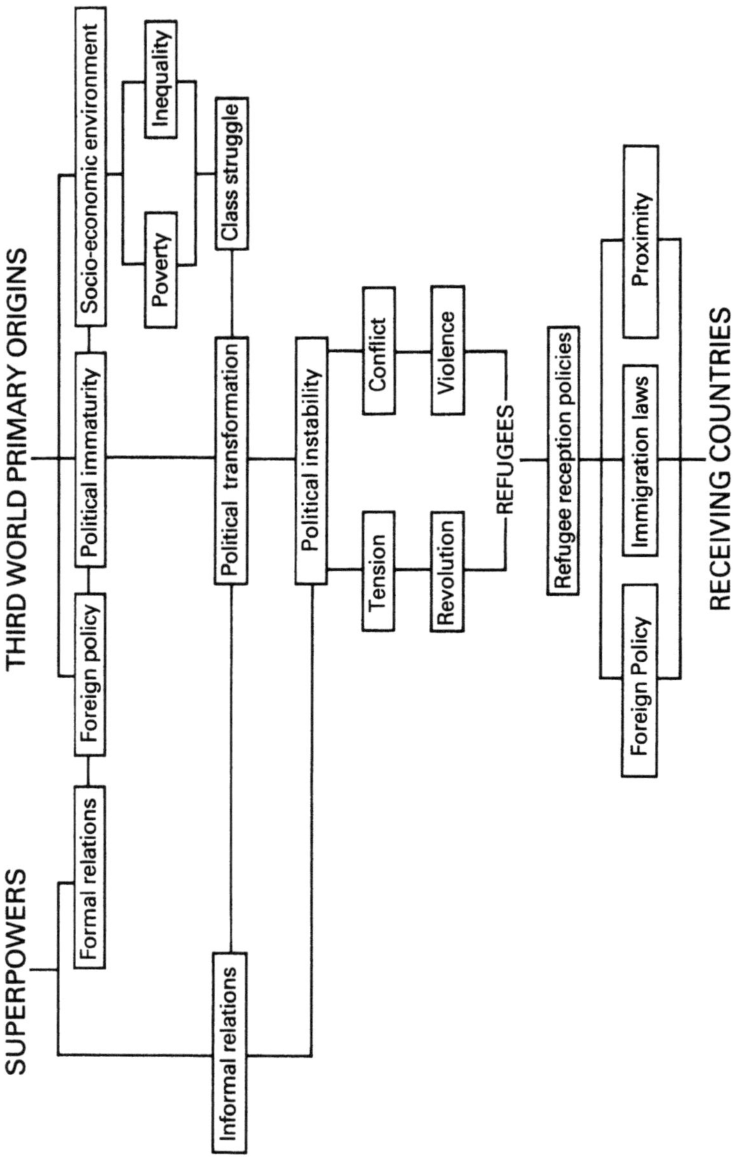

Fig. 6.2 A global refugee systam

countries; the reception countries; and the super and major regional powers.

Most refugee flows stem from Third World countries, many recently independent, and characterised by political immaturity. Most of these countries do not have democratic forms of government, nor well established political systems. As states they are weak. This immaturity fosters the political transformation discussed by Zolberg; its endemic nature leads to instability, tension and conflict from which flow revolution and violence. Such upheavals inevitably lead to outflows of some people - refugees. These states, weak politically, are also weak socially and economically with endemic poverty and inequality. Power struggles occur between social classes and groups and engender political transformations. These internal conditions may be exacerbated by the actions of major regional and super powers. At a formal foreign policy level influence may be exerted in an overt way, bringing countries clearly within the orbit of the major power. As Zolberg has indicated, even the poorest states acquire some value as pawns in great power struggles. Major powers may also be involved in informal, often covert action to destabilise, seeking to bring about political transformations and foster instability directly. There are numerous Third World countries that lie on the fringes of super power influence spheres, and where conflict has become endemic. Instability and conflict, bred from the interplay of internal and international forces, combine to cause refugee flows.

The recognition of people as refugees depends upon the attitudes of potential receiving countries. Zolberg has pointed out that the refugee policies of potential receivers are often an adjunct of foreign policy. States will more readily accept as refugees persons fearing violence at the hands of their antagonists than of their friends. Hence comparable events may trigger flows in one case because there is somewhere to go, but not in another because there is not.

Receiving countries are, therefore, the third major political element in the system. The highly politicised definition of a refugee means that the legal practices adopted by states become an essential part of the functioning of the system. These become codified into a refugee reception policy which has two major influences in addition to foreign policy. First are the immigration laws that have been developed and which determine the degree of openness of borders to the nationals of other countries. The granting of asylum may be part of such laws. Second, geographical proximity is a major influence. Most refugees go at first to a neighbouring state under the gravity principle. They may congregate there or move to a more permanent exile further away.

Quite clearly a lot of hypothesis testing needs to be done on this model, but preliminary indications are that such a theoretical framework, stressing as it does both internal and external relations of states, provides a good starting point for analysing flows. Since refugees are politically motivated for the most part, it is in political relationships that causation is most likely to lie. The geographical interest must be in studying refugees as a manifestation of political systems that display spatially unique characteristics. Just as it cannot be assumed that global economic forces explain all labour migration patterns, so we must expect similar political forces to result in geographically different types of refugee flows.

As yet there has been little attempt by geographers to demonstrate how the interplay of macro forces with local situations produces a geography of refugees. However, there is plenty of evidence that variations in the global system described above produce spatially varying refugee responses. This can be illustrated by looking at contrasts in refugee flows to selected countries in both the Third World and in Europe.

Pakistan is currently host to the largest refugee population in the world (Figure 6.1), nearly three million. Most of these are from Afghanistan and live in camps near the border, but in addition there are two hundred thousand refugees from Bangladesh and about two thousand Iranians (Crisp and Nettleton, 1984). These three origins are associated with three very different political transformations - the first caused by foreign invasion, the second mainly by racial and religious differences, the third by a combination of civil and foreign war, and violent political upheaval. The Afghans live in three hundred and forty village camp sites, the largest with thirty thousand refugees, mainly in Baluchistan. About three quarters of the refugees are women and children, the menfolk remaining behind, many to fight. Aid in the camps comes jointly from the Pakistan government, which assigns refugees to camps and gives them a minimum subsistence allowance, and from international aid organisations. The policy is mainly to promote self-sufficiency, with resettlement in third countries not yet under consideration. In this situation the refugees are clearly regarded as a potentially influential factor in the political situation in Afghanistan.

A rather different situation exists in Thailand, another major destination for local refugee flows, where since 1975 over six hundred thousand refugees have come in from Kampuchea, Laos and Vietnam. Unlike the policy of the Pakistan government, that of Thailand is to resettle as many refugees as possible in third countries, reflecting the political difficulties of cross-border conflicts between Khymer Rouge soldiers who infiltrate and use refugee camps, and the pursuing Vietnamese, with the Thai population and army caught in the crossfire. Consequently, Thai policy is to deter new arrivals by long term detention in austere camps and with limited access to resettlement processing facilities. Clearly, unlike Pakistan where reception policies are likely to encourage immigration of refugees, in Thailand the reverse is the case.

In contrast to both of these are the policies developed to deal with refugees by European countries, since the end of World War II and the descent of the Iron Curtain no longer in the front line (with the possible exception of France and the movement there of pieds noirs in 1962-3). Reception countries in Western Europe, like Sweden and Switzerland, receive their refugees largely as third countries and make distinctions between quota refugees and individual asylum seekers. Reception is restricted but, in contrast to countries in close geographical proximity to primary sources, well-organised with integration being the principal aim.

CONCLUSION

There is no coherent spatial theory of international migration. The cement of space is distance, and this has proved too friable to cope with the stresses of human behaviour. Distance is important, of course, but only in the way it makes us look for the significance of distance-related phenomena like information flows and cultural proximity. Gravity-based theories can be, at best, starting points in the quest for explanation. Regression models can be helpful, especially at local and regional levels, in pointing out associations of variables, but the theoretical links between them reflect human attitudes and behaviour. Too often such models assume homogeneity among potential migrants, and they are usually static in structure. Taken as descriptive rather than explanatory models, however, gravity and regression formulations are capable of allowing some predictions to be made with reasonable accuracy over the short term.

Ultimately the geographical approach to international migration is both empirical and pragmatic. Its standpoint is spatial variability and for that it requires empirical data, too often lacking. Where empirical data have been available they have too often shown that deductive theories do not apply. The pragmatism stems from a concern with real world problems. Migration is regarded as one process of spatial change that will interact with others to produce regional differentiation. Flow patterns are, therefore, regarded as one component in an interacting system of places.

It is in its diffuse nature that the strengths and weaknesses of the spatial approach may be discerned. By emphasising diversity and variability over space and time within and between migration networks it alerts us to the pitfalls of the general approach. But in focussing on differentiation it hinders progress towards general theory. At the end of the day, perhaps its ability to show how different theoretical approaches may be combined in particular local, regional and national circumstances can help us spell out the limits within which we might aspire to a general theory.

REFERENCES

Abella, M.I. (1984) 'Labour Migration From South and S.E. Asia: Some Policy Issues', International Labour Review 123 (4), 491-506

Augarde, J. and Prevost, G.(1970) 'La Migration Algerienne', Hommes et Migrations: Etudes, 116, 3-161

Bach, R.L. (1983) 'Emigration From the Spanish-Speaking Caribbean', in M.M. Kritz (ed.), U.S. Immigration and Refugee Policy, Lexington

Barber, M. (1984) 'Refugees and Superpower Rivaly', in J. Crisp and C. Nettleton (eds.), Refugee Report 1984, British Refugee Council, London

Baucic, I. (1977) 'Regional Differences in Yugoslav External Migration', in H.L. Kostanick (ed.), Population and Migration Trends in Eastern Europe, Boulder, Colorado

Birks, J.S. and Sinclair, C.A. (1980) International Migration and Development in the Arab Region, I.L.O., Geneva

Bohning, W.R. (1970) 'The Differential Strength of Demand and Wage Factors in Intra-European Labour Mobility: with Special Reference to West Germany, 1957-68', International Migration 8, (4), 193-202

Bohning, W.R. (ed.) (1981) Black Migration to South Africa, I.L.O., Geneva

Crisp, J. and Nettleton, C. (1984) Refugee Report 1984, British Refugee Council, London

Drettakis, E.G. (1975) Yugoslav Migration to and from West Germany, 1962-1973, University of Zagreb, Centre for Migration Studies

Findlay, A. and Godden, J. (1985) 'U.K. Expatriates in the Middle East', Middle East Economic Digest, January, 31

Findlay, A. and Samha, M. (1985) 'Return Migration and Urban Development: a Jordanian Case Study. Paper Presented at Return Migration Symposium', I.B.G. Conference, Leeds

Gleave, D. (1983) The Relationship Between Labour Mobility and Technical Change, Technical Change Centre, London

Guha, A. (1978) 'Intra-COMECON manpower Migration', International Migration 14, 52-65

Jennings, E.E. (1971) Routes to the Executive Suite, New York

Jones, F.L. (1964) 'The Territorial Composition of Italian Emigration to Australia: 1976-162, International Migration, 2, 247-265

Jones, H.R. (1973) 'The Regional Origin of Emigrants: Findings from Malta', International Migration, 11, 52-65

Keely, C.B. and Elwell, P.J. (1981) 'International Migration: Canada and the United States, in M.M. Kritz, C.B. Keely and S.M. Tomasi (eds.), Global Trends in Migration, New York

King, R. and Strachan, A. (1979) 'Spatial Variations in Sicilian Migrations: a Stepwise Multiple Regression Analysis', Mediterranean Studies, 2, (1)

King, R. and Strachan, A. (1980) 'Patterns of Sardinian Migration', Tijdschrift Voor Economische en Sociale Geografie, 71 (4), 209-222

King, R., Strachan, A., and Mortimer, J. (1985) 'Return Migration and the Development of Southern Italy. Paper Presented at Return Migration Symposium', I.B.G. Conference, Leeds

Kritz, M. (1981) 'International Migration Patterns in the Caribbean Basin: An Overview', in M.M. Kritz, C.R. Keely and S.M. Tomasi, (eds.)

Kunz, E.D. (1973) 'The Refugee in Flight: Kinetic Models and Forms of Displacement', International Migration Review, 7, (2), 125-146

Kunz, E.F. (1981) 'Exile and Resettlement: Refugee Theory', International Migration, Review, 15, 42-51

Lawless, R.I. and Seccombe, I.J. (1984) 'North African Labour Migration: The Search for Alternatives. Paper presented at I.B.G. Conference, Durham

Lee, E.S. (1967) 'A Theory of Migration', Demography, 3, 47-57

Lewis, J.R. and Williams, A.M. (1985) 'The Economic Impact of Returned European Migrants and Retornados in central Portugal.' Paper presented at return migration symposium, I.B.G. conference, Leeds

Lianos, T.P., (1975) 'Flows of Greek out-Migration and Return Migration', International Migration, 13, 119-133

Ling, L.H-M. (1984) 'East Asian Migration to the Middle East: Causes, Consequences and Considerations', International Migration Review 18 (1), 19

Mabugunje, A.L. (1970) 'Systems Approach to a Theory of Rural-Urban Migration', Geographical Analysis, 2, 1-18

McDonald, J.R. (1969) 'Labour Immigration in France, 1946-1965', Annals of the Association of American Geographers, 59, 116-134

Marshall, A. (1981) 'Structural Trends in International Labour Migration: the Southern Cone of L. America', in M.M. Kritz, C.B. Keely, and S.M. Tomasi, (eds.)

Morales, R. (1983) 'Transitional Labour, Undocumented Workers in the Los Angeles Automobile Industry', International Migration Review 17, (4), 579-596

Parenti, G. (1958) 'Italy', in B. Thomas (ed.), Economics of International Migration. New York

Petersen W. (1958) 'A General Typology of Migration', American Sociological Review, 23 (3), 256-266

Portes, A. (1981) 'Modes of Structural Incorporation and Present Theories of Labour Immigration', in M.M. Kritz, C.B. Keely, and S.M. Tomasi, (eds.)
Ranney, S. and Kossoudji, (1983) 'Profiles of Temporary Mexican Labour Migrants to the U.S.', Population and Development Review, 9(3), 475-493
Ravenstein, E. (1885) 'The Laws of Migration', Journal of the Royal Statistical Society, 48, 167-227
Royal Scientific Society, (1983) Worker Migration Abroad, R.S.S., Amman, Jordan
Salt, J. (1981), 'International Labour Migration in Western Europe: A Geographical Review', in M.M. Kritz, C.B. Keely, and S.M. Tomasi (eds.)
Salt, J. (1983/4) 'High Level Manpower Movements in North-west Europe', International Migration Review, 17, 633-652
Salt, J. (1985) 'West German Dilemma: Little Turks or Young Germans?' Geography, 70(2), 162-168
Saraceno, E. (1985) 'The Professional Resettlement of Return Migrants and Regional Development: The Case of Friuli-Venezia Giulia. Paper presented at Return Migration Symposium', IBG Conference, Leeds
Schulz, P. (1975) 'Labour Migration Among the Socialist European Countries in the Post-W.W.II period', International Migration, 11, 190-201
Secombe, I.J. and Lawless, R.J. (1985) 'Some New Trends in Mediterranean Labour Migration: The Middle East Connection', International Migration, 23(1), 123-148
Simon, G. and Noin, D. (1972) 'La Migration Maghrebine vers l'Europe', Cahiers d'Outre-Mer, 25, 241-276
Stahl, C.W. (1982) 'International Labour Migration and International Development. I.L.O. International Migration for Employment', WPI.
Stahl, C.W. (1984) 'Singapore's Foreign Workforce: some Reflections on its Benefits and Costs', International Migration Review 18(1), 37-49
Stein, B.N. and Tomasi, S.M. (1981) 'Refugees Today', International Migration Review, 15. Special issue
Tapinos, G.P. (1966) 'Migrations et particularismes regionaux en Espagne', Population, 21, 1135-1164
United Nations, (1982) 'International Migration Polices and Programmes: A World Survey', Population Studies No. 80, United Nations, New York
Wadensjo, E. (1978) 'The Economic Determinants of Migration in Countries of Destination', paper in Conference on Economic and Demographic Change: Issues for the 1980s. I.U.S.S.P. Liege
Wolpert, J. (1965) 'Behavioural Aspects of the Decision to Migrate', Papers of the Regional Science Association, 15, 159-169
Zolberg A. (1985) 'Political Forces as Determinants of International Migration. Paper presented at International

Union for the Scientific Study of Population', Seminar on 'Emerging Issues in International Migration, Bellagio
Zubrzycki, J. (1981) 'International Migration in Australasia and the South Pacific', in M.M. Kritz, C.B. Keely and S. M. Tomasi (eds.)

Chapter Seven

INTERNAL MIGRATION IN THE THIRD WORLD*

J. Gugler

This chapter, in contrast to the other contributions to this volume, has an explicit regional focus on the less developed countries of Asia, Oceania, Africa, Latin America, and the Caribbean. These countries are poor by definition. The majority of people typically live in rural areas where most continue to farm with limited inputs. And even though urban growth exceeds natural population growth in every country, the rural population continues to grow almost everywhere. Still, there is considerable variation among Third World countries: income levels range widely; and while some countries are only beginning to urbanise, others have nearly completed the urban transition (Table 7.1). Within the Third World, Latin America constitutes a distinct region further along in both, economic development and urbanisation.

The largely rural character of most Third World countries makes two migratory patterns stand out: rural-rural migration and rural-urban migration. Large numbers of people participate in rural-rural migration in many Third World countries. They usually move to more abundant or more fertile lands. Natural calamities, such as the drought in the Sahel zone in Africa, precipitate such movements. Public agencies may assist migrants or indeed recruit them. The 'transmigration' scheme in Indonesia to move people from the most densely populated islands - Java, Bali, and Lombok - to other islands, is the most ambitious such project. About 2.5 million people have been moved between 1979 and 1984, and government plans call for 65 million people to be shifted up to 3,000 kilometers from their present homes over the next 20 years. When resettlement is successful, rural-rural migrants, unlike their cousins who go to the city. are assured of

* This chapter draws on my discussion of rural-urban migration and urban employment in Gilbert and Gugler (1982). I wish to thank, without implicating, William G. Flanagan for helpful comments, Saskia Gugler for editing.

(self-)employment and of an economic asset, their land, that can sustain them in old age. Hence they are therefore less likely to invest in maintaining a claim to land in their area of origin. As in the city, however, they may face hostility from indigeneous people or other immigrant groups.

THE URBAN TRANSITION

Rural-urban migration in the Third World commands special attention. Migrants put strain on urban labour markets characterised by widespread unemployment and underemployment. They escalate demand for urban housing and services already considered inadequate. For better or worse, they exacerbate popular pressures on the political system.

Rural-urban migration in the Third World also arouses interest because, in the second half of the 20th century, it constitutes the last phase in a great human revolution: it completes the urban transition. Urban settlements were first established more than five thousand years ago; but as recently as the beginning of this century, only one in eight people lived in an urban area. Before the end of the century, half of humanity, three billion people, will live in urban settlements; and two thirds of that number will live in the Third World.

The magnitude of this last phase of the urban transition, the sheer number of people involved in it, is without precedent in human history. Urban growth is proceeding at a rapid rate in all Third World countries. And while it has slowed down in many countries, indeed throughout Latin America, it has accelerated in others, especially in South Asia and Subsaharan Africa. Countries such as Bangladesh, Cameroon, Kenya, Mozambique, Nepal, Saudi Arabia, Tanzania, report that their urban populations more than doubled between 1973 and 1983 (Table 7.1). The largest Third World cities are about to outstrip their First World counterparts: Mexico City and São Paulo are projected to become the world's largest cities before the turn of the century (United Nations 1980:58). The rate of urbanisation, i.e. the increase in the proportion of the population urban, in the Third World now, however, is similar to the rate that characterised the urban transition in Europe 75 years earlier. In the quarter-century from 1950 to 1975, the proportion urban grew from 16.7% to 28.0% in Third World countries; in today's more developed countries, it grew from 17.2% to 26.1% between 1875 and 1900. The slight difference in the two rates of urbanisation is well within the margin of error of the estimates.

While the rate of urbanisation in the Third World is not exceptionally rapid by historical standards, the rate of growth of its urban population has no parallel. Between 1875

Table 7.1: Demographic Characteristics and Income of Major Third World Countries[1]

Region/country	Population (millions) mid-1983	GNP per capita (dollars) 1983[2]	Urban population as percentage of total population 1983[3]	Urban sex ratio (males per 1,000 females) circa 1970	Urban growth, average annual growth rate (percent) 1965-73	Urban growth, average annual growth rate (percent) 1973-83	Population growth, average annual growth rate (percent) 1973-83
East Asia							
Burma	35	180	29	1040	4.0	3.9	2.0
China	1019	300	21	na	na	na	1.5
Indonesia	156	560	24	1000	4.1	4.8	2.3
Korea, North	19	(930)	62	na	4.9	4.2	2.5
Korea, South	40	2010	62	996	6.5	4.8	1.6
Malaysia	15	1860	31	1013	3.3	3.5	2.4
Philippines	52	760	39	933	4.0	3.8	2.7
Thailand	49	820	18	983	4.8	3.6	2.3
Vietnam	59	(240)	20	na	5.5	2.4	2.7
South Asia							
Bangladesh	95	130	17	1294	6.6	7.6	2.4
India	733	260	24	1166	4.0	4.2	2.3
Nepal	16	160	7	1173	4.3	8.2	2.6
Pakistan	90	390	29	1227	4.3	4.3	3.0
Sri Lanka	15	330	26	1133	3.4	2.9	1.7

Middle East and North Africa							
Afghanistan	17	(230)	17	na	5.6	6.2	2.6
Algeria	21	2320	46	999	2.5	5.4	3.1
Egypt	45	700	45	1054	3.0	2.9	2.5
Iran	43	(1690)	53	1085	5.4	5.1	3.1
Iraq	15	(1930)	69	1029	5.7	5.3	3.6
Morocco	21	760	43	958	4.0	4.2	2.6
Saudi Arabia	10	12230	71	na	8.4	7.4	4.7
Syria	10	1760	48	1071	4.8	4.2	3.3
Turkey	47	1240	45	1148	4.9	3.7	2.2
Subsaharan Africa							
Cameroon	10	820	39	na	7.3	8.4	3.1
Ethiopia	41	120	15	903	7.4	6.0	2.7
Ghana	13	310	38	996	4.5	5.3	3.1
Kenya	19	340	17	1386	7.3	8.0	4.0
Mozambique	13	(370)	17	na	8.2	10.2	2.6
Nigeria	94	770	22	1149	4.7	5.1	2.7
South Africa	32	2490	55	1119	2.6	3.9	2.4
Sudan	21	400	20	na	6.3	5.5	3.2
Tanzania	21	240	14	1078	8.1	8.6	3.3
Uganda	14	220	7	1191	8.3	0.3	2.8
Zaire	30	170	38	na	5.9	6.9	2.5

Table 7.1: (continued)

Region/country	Population (millions) mid-1983	GNP per capita (dollars) 1983[2]	Urban population as percentage of total population 1983[3]	Urban sex ratio (males per 1,000 females) circa 1970	Urban growth, average annual growth rate (percent) 1965-73	Urban growth, average annual growth rate (percent) 1973-83	Population growth, average annual growth rate (percent) 1973-83
Latin America							
Argentina	30	2070	84	953	2.1	2.1	1.6
Brazil	130	1880	71	939	4.5	4.1	2.3
Chile	12	1870	82	906	2.8	2.4	1.7
Colombia	28	1430	66	868	4.4	2.9	1.9
Cuba	10	(1180)	70	983	2.8	1.9	0.8
Mexico	75	2240	69	966	4.8	4.1	2.9
Peru	18	1040	67	1000	4.7	3.6	2.4
Venezuela	17	3840	85	961	4.8	4.3	3.5

na - Not available

1. This table includes all Third World countries with a population over 9.5 million in mid-1983.
2. GNP per capita figures (in brackets) are for 1982 and from a different source; they are not strictly comparable with the other GNP per capita figures.
3. Percentages urban are based on different national definitions of what is 'urban', and cross-country comparisons should be interpreted with caution.

Source: All data from World Bank (1985: annex tables 1, 19 and 22), except GNP per capita 1982 (in brackets) from Sivard (1985: statistical annex table III), urban sex ratios from Ferree and Gugler (1985).

and 1900, the urban population in the now more developed countries grew by 100% and the rural population by 18%. While Third World countries were traversing roughly the same range in proportion urban between 1950 and 1975, their urban population grew by 188%, their rural by 49%. The difference in growth rates between the two eras is explained by much higher rates of natural population growth in Third World countries today. Urban growth is exceptionally rapid in developing countries, not because of an unusually rapid increase in the proportion of the population urban produced by rural-urban migration, but because of the fast increase in the total population to which this proportion is applied (Preston 1979).

Until the last century, many rural populations had little or no connection with urban centres. They lived in quite self-centred societies. By and large, they operated subsistence economies, and maintained only limited external contracts. The expansion of the capitalist system, under way for half a millenium, and accelerated by the Industrial Revolution, has however incorporated ever more outlying regions into the emerging world economy. By now the process is virtually complete. The world over, rural populations have been drawn into the urban nexus. Subsistence farmers have come to produce for urban markets, either self-employed as peasants, or on the account of land owners. They have become subject to political control exerted from urban centres. And their culture has been transformed as they have been indoctrinated by foreign missionaries, subjected to school curricula mandated by urban elites, and exposed to films, and radio and television programs, produced in distant cities, some of them half way across the globe.[1]

For many peoples the integration into the world capitalist system was traumatic. They were enslaved, were forced into the encomienda, or suffered as indentured labour. They were thus tied to the land even while the product of their labour entered world markets. But such bonds have largely been broken. Today rural populations raise the cash they need to settle taxes, to purchase manufactured goods, and to pay school fees, by selling some of the products they grow as independent producers, as tenant farmers, and as share croppers. Or they find employment with local landowners, in other rural areas, or in the cities. Whether they sell produce on the market or their labour to an employer, they are part of a far-reaching economic system beyond their control. They experience the vagaries of the world economy when they lose

---

[1]For analyses of the incorporation of rural populations, see Pearse (1970) on Latin America, and Gugler and Flanagan (1978a:180-3) on West Africa.

their jobs during recessions or when a sudden drop in world market prices depresses their earnings from export crops.

Some rural populations were exploited to such an extent that their living conditions declined. Elsewhere, specific groups experienced pauperization. But for most rural dwellers living conditions have improved; in terms of better health and longer life, the change has usually been dramatic. However, these very improvements have accelerated population growth, and population pressure on the available land has become severe in many areas.

Incorporation into the world system brought new elements of differentiation to rural people - many of whom had known a measure of equality based on general access to land. The first men to be converted by missionaries and to attend their schools had a headstart in employment as teachers, government officials, or commercial clerks; the first to accumulate a little capital in employment, or from the sale of their crops, established themselves as traders or transporters; and the first to become agents for the colonial government or the independent state expanded their control over land, or derived the benefits that flow from wielding patronage. Rural populations thus came to experience relative deprivation. As incorporation proceeded, they recognised their own poverty: they saw a few in their midst rise to levels of affluence undreamt of in the past, and they came face to face with the lifestyle of outsiders - missionaries, traders, government officials, foreign experts, and tourists.

With the perception of a better life enjoyed by some, came an awareness of the means towards such an end. Throughout the world today, most rural dwellers know what it is like to sell and buy in markets or shops, they have seen what a school certificate can do for the future of a child, have listened to first-hand accounts of those who work in the city. Some improve their condition while staying in rural areas, or by moving to other rural areas, as farmers, traders, or artisans. But for many, rural prospects appear dim and the urban scene more promising.

THE RURAL-URBAN GAP

Comparisons of rural and urban incomes are notoriously problematic.[2] The usual basis for comparison is an index of urban wage rates and a crude index of agriculture incomes,

---

[2]Lipton (1977:146-53, 430-34) provides data on rural-urban differentials in wages, incomes, and expenditures for nineteen developing countries and discusses the shortcomings of such data.

such as the prices which cash crops fetch. These measures signal sudden shifts in urban or rural incomes, e.g. the rapid rise in urban wages in most African countries just before and shortly after the attainment of independence. But they have serious shortcomings. Adjustments for the difference in the cost of living between rural and urban areas are difficult to make. And producer prices fail to take into account changes in the productivity of labour in agriculture, which may be decreased by growing population pressure, but increased by improvements in farming techniques.

The foremost difficulty for any comparison of urban and rural incomes intended to explain and predict migratory behaviour is the need for disaggregation. Average urban wage rates have little relevance for the unskilled migrant. Rural opportunities will vary according to the endowment of the region of origin and the migrant's local position, i.e. access to land and capital inputs. The point is well illustrated by Shaw's (1976:74-105) analysis of rural out-migration within and among Latin American countries. Variations in the system of land tenure are significantly related to the rate of rural out-migration from different provinces in both Chile and Peru. Comparing sixteen countries, Shaw shows further that it is not absolute population pressure, but access of labour to land as mediated by land ownership, that affects rural out-migration. The average rate of out-migration was highest for those countries, Mexico, Peru, Venezuela, in which more than half the land was held by latifundios and where more than half the farms were minifundios.

An evaluation of the rural-urban balance of opportunities has to take into account not only individual incomes but also public amenities. The urban areas, and especially the major cities, invariably have more and better facilities than their rural hinterlands, and offer superior education and training, for the migrant's children in particular. Expert medical care and drugs can be found. Piped water assures clean water and releases women from the drudgery of fetching water over long distances. Electricity replaces the kerosene lamp and the open fire. Some migrants eventually succeed in obtaining subsidised housing. Clearly, here again, there is a serious problem of disaggregation. Migrants frequently experience severe discrimination in access to these urban amenities, indeed, for some, housing and sanitary conditions may well be worse than where they came from. Still, on balance most enjoy more amenities than those who stayed behind.

Rural-urban differences are best assessed with an indicator that measures well-being directly. Health and life are universal values and thus provide measures with cross-cultural validity. Mortality data are an obvious indicator. Estimates of infant and early childhood mortality broken down by rural versus urban residence have been assembled recently for 44 Third World countries (Gilbert and Gugler

1982:53). In all but one, rural mortality was reported higher than urban; in 24 countries the rural rate exceeded the urban by more than 30 percent. The reliability of many of these data is questionable, but there is reason to assume that they tend to understate rather than exaggerate urban-rural differentials. Information on infant and child mortality is usually derived from retrospective surveys. Characterisation as urban or rural depends on residence at the time of the survey. Thus some of the births and deaths reported as urban occurred in rural areas, before the migration of the mother, and to that extent high rural mortality rates increase the urban average. And the major source of error, under-reporting of children who have died, is more common in rural than in urban areas. Still, the problem of disaggregation remains. At the urban end in particular, mortality rates among migrants presumably are above average. The difference should not be exaggerated though: migrants usually constitute a large proportion of the urban population and hence strongly affect urban averages. In most countries the rural-urban differential is sufficiently large to suggest a real improvement in living conditions for migrants.

Cities are centres of power and privilege - such a summary statement holds throughout the Third World today. Even socialist countries, while reducing income inequality within the urban and rural sectors respectively, appear to find it difficult to deal with inter-sector inequality. In China, personal consumption has been estimated at $244 per capita in urban as against $111 per capita in rural areas in 1979. The ratio of 2.2 is similar to that of other developing countries in Asia. Despite a commitment to reducing the urban-rural gap, urban per capita incomes are estimated to have increased between 1957 and 1979 at an annual average real rate of 2.9 per cent, but rural incomes at only 1.6 per cent. The gap in personal income is accompanied by a large gap in collective consumption. The quality of educational and health facilities in particular is much higher in urban areas (World Bank 1981:52-7; see also Nolan and White 1984).

Nyerere, the President of Tanzania, focused attention on the issue a few years after his country had gained independence. In the famous Arusha Declaration (Nyerere [1967] 1968:28) he warned:

> Although when we talk of exploitation we usually think of capitalists, we should not forget that there are many fish in the sea. They eat each other. The large ones eat the small ones, and small ones eat those who are even smaller. There are two possible ways of dividing the people in our country. We can put the capitalists and feudalists on one side, and the farmers and workers on the other. But we can also divide the people into urban dwellers on one

> side and those who live in the rural areas on the other. If we are not careful we might get to the position where the real exploitation in Tanzania is that of the town dwellers exploiting the peasants.

Lipton (1977; see also Lipton 1984) diagnoses the problem as 'urban bias'. He argues that rural people in the Third World are provided with fewer resources than is either equitable or efficient because allocations are urban-biased. At issue is not dishonesty or errors of analysis, but a prevailing disposition of policy makers, and their advisors, to justify heavily urban resource allocations. Lipton offers a major demonstration of the inequity and inefficiency of these allocations. His explanation of the dispositional bias focuses on the fact that decisions are made in the city. In taxation policy, in the allocation of investment resources, in the pricing of rural products, in the provision of industrial and agricultural inputs, in education and research: everywhere it is government by the city, from the city, for the city.

Dispositional bias in favour of the city is certainly common among Third World elites, but political power relations need also to be taken into account. After all, international agencies and foreign donors have preached the importance of rural development for many years - to little effect. Government leaders are preoccupied with maintaining the support of the groups that condition their chances of staying in power: the armed forces, investors, professionals, strategically placed elements of labour - all of them urban-based. Even the lumpenproletariat poses the threat of urban riots. Resources for both consumption and investment are disproportionately allocated to cities because the rural masses have little political leverage. In most Third World countries they have been disenfranchised. Where they can participate in competitive elections their vote is frequently controlled by local elites through patronage or outright coercion. Peasant rebellions tend to be isolated and rarely constitute an effective threat to urban elites, and boycotts of purchasing agencies are similarly difficult to organise across the countryside. The rural masses are thus usually without a voice, and their fate is neglect. They are left with the option to vote with their feet.

WHY PEOPLE MOVE

A substantial body of research on rural-urban migration has been accumulated over the last two decades, and the evidence is overwhelming: most people move for economic reasons. When migrants are asked, why they moved, they usually give the better prospects in the urban economy as their chief reason.[3] Migration streams between regions have been shown to correspond to income differentials between those regions.[4] And over time, as economic conditions at alternative destinations change, migration streams switch accordingly. Rural-urban migration constitutes the attempt to join an environment that offers better life chances in the immediate future, over a working life, for the migrant's descendants.

The sight of severe and widespread poverty in Third World cities easily leads to the assumption that migrants did not know what to expect, that illusions about the prospects lying ahead brought them to an urban environment in which they are trapped. In fact, however, most migrants are quite well informed before they move. Many had heard the accounts of earlier migrants who had returned to the village on a visit or to stay. Some had been able to visit kin or friends in the city before deciding to move. Studies throughout the Third World report, time and again, that most migrants consider that they have improved their condition, and that they are satisfied with their move.

Migration is rarely a solitary affair. Even when individuals migrate alone, others typically assist them in the move, in adapting to the urban environment, in securing a foothold in the urban economy. The extended family usually constitutes the central element in such a network. Relatives are most likely to help to pay for an education that will prepare the future migrant, to provide a home for children who have been sent to town to pursue their education, to offer the newly arrived migrant shelter and food for a while. Once migrants are established in town, they can thus generally be expected to attract new migrants from their home area, to initiate chain migration.

Chain migration encourages direct moves, even over large distances, in contrast to "stage migration," i.e. migrants moving to one or several intermediary destination(s) before reaching their final destination. There is intuitive appeal to the notion that rural emigrants go first to small towns and, after a period of adaptation to the urban environment, move on to metropolitan areas. Or the concept of stage migration may be stretched to span two generations: the first makes it to the local town, the second into the city. At present, however, we lack the data to substantiate the

---

[3]For an extensive review of the literature on internal migration in Africa, Asia, and Latin America, see Simmons, Diaz-Briquets, and Laquian (1977); for a general review of the migration literature that focuses on theory construction, Shaw (1975); for a comprehensive bibliography of literature on labour migration in Latin America, Lowder (1978).

[4]Yap (1977) provides a critical review of the econometric studies that have come to dominate research since the late 1960s.

view that stage migration is a significant mode of rural-urban migration in Africa, Asia, or Latin America (Simmons et al., 1977:29,58,95).

If migration is embedded in social relations, the decision to migrate, also, is not arrived at in isolation. Groups, such as family, local community, classmates, evolve patterns of behaviour which are modified over time as experience dictates. Gone are the days when elders disapproved of young men 'running away', forcing them literally to abscond in the night. Today, migration is accepted virtually everywhere, and frequently the remittances of migrants furnish villagers with what for them are the luxuries of life. In some cases, and notably in Southern Africa, those who stay in the rural areas have become dependent for their livelihood on the earnings of those who have found urban employment.

Once a pattern of migration is established, going to town may become the thing to do, the urban experience takes on positive connotations. Thus, in many parts of the Sudanic belt, young men were expected to have spent one or several spells of seasonal migration in Ghana (Rouch, 1956:194). Such norms usually emphasise the experience to be gained, and the challenge to the young to prove themselves. Sometimes these norms become so generally accepted that individuals are swept along, even when they do not share the economic rationale for going to the city. Migration may, in this way, become the rule for a community. Nonetheless, once economic circumstances change, collective migration behaviour quickly adapts to the new reality.

The push from rural areas, and the pull of urban areas, are often distinguished in discussions of migration. Reference to push or pull serves to emphasise the importance of a particular motive in the decision to migrate. Refugees, for example, may be said to be pushed out of their rural homes. During the civil strife that accompanied the Partition of India in 1947, about 16 million people fled across the newly established boundaries; most of those uprooted from rural areas sought a new beginning in cities. War, the man-made calamity, frequently makes rural areas so insecure that peasants pack up and leave for the relative security of cities. During the many years of the blood bath in Indochina, some peasants sought shelter in the cities, though many others were relocated by force. In Indonesia, the independence struggle, as well as regional rebellions after independence, led to a mass exodus from the affected rural areas. Civil wars in Korea, Malaysia, the Southern Sudan, and Zaïre made peasants abandon their ancestral lands. In Colombia, La Violencia, the violent conflict in the countryside which lasted for over a decade, was a major factor in rural-urban migration according to some observers.

Elsewhere droughts, earthquakes, cyclones, volcanoes, or floods have brought immediate physical danger, as well as threatening hunger and disease. Such disasters frequently make rural dwellers abandon their homes and seek relief in urban areas. They are commonly referred to as 'natural', but they are man-made to the extent that political action, or inaction, increases their severity and impact on the affected population. Thus the famines of the 1970s and 1980s in Subsaharan Africa were not the necessary outcome of an act of nature. They must be traced first and foremost to the policies of African governments that did little to develop rural areas and left much of the peasantry in a precarious condition, without the resources to deal with a natural calamity. Not only was the peasantry disadvantaged in general, but there were few specific efforts to make marginal lands less vulnerable to drought, e.g. by sponsoring well construction, reforestation. Indeed, many rural areas lack the roads that could carry relief supplies: the hungry have no alternative but to move to urban areas or camps. The disregard of urban decision makers for the rural masses was exposed when some governments refused to acknowledge famine conditions, and relief operations were delayed as a consequence.

The push from rural areas is dramatised in the case of refugees. Even in such extreme cases, however, it can be seen that a comparison is involved: the refugees move to more secure settings. Invariably, the decision to migrate involves an assessment of alternative locations; people move to a more promising environment. This is also the case where the pull looms large. To take the archetypal case, joining the Gold Rush implied a perception of greater opportunities Westward.

Today, great masses of rural people are potentially mobile. And they appreciate the gap between rural and urban standards of living. Sometimes they are said to be pushed out by rural poverty, or to be attracted by urban opportunities. They are more accurately seen as comparing their prospects in the rural as against the urban setting, weighing their life chances in the two environments.

THE URBAN LABOUR MARKET

There have been times when employers in the Third World clamoured for workers, when they complained about high rates of turnover and widespread absenteeism. In the early stages of the incorporation of the rural economy, subsistence farmers saw little reason to sell their labour elsewhere. After the abolition of slavery, colonial governments resorted to forced labour. Then in the imposition of taxes a more subtle means of coercion was found: unless they grew cash crops or parted with some of their cattle, people were forced to earn wages to pay their taxes. Nearly everywhere, however, incorporation soon created new demands in the rural areas that only money could satisfy. The high degree of self-sufficiency of the traditional farmer was compromised as he became dependent on goods and services purchased in the market. Some raised the cash within the rural economy, but many went to work on plantations, in mines, and in the cities.

Even where rural areas were effectively incorporated, labour shortages were common, and the urban work force was frequently described as uncommitted. The level of urban wages, and working and living conditions, were at issue. Thus, a cheap-labour policy characterised much of colonial Africa. The policy was buttressed by the proposition that migrants would work less at higher wage rates because they would stay in town only as long as necessary to meet a fixed objective, that they were target workers. In most settings, the proposition that the labour-supply function was thus backward-sloping bore little relationship to reality; rather, it was a myth that provided the ideological underpinning for a cheap-labour policy (Berg 1961).

Labour shortages are a thing of the past. For years, urban labour markets throughout the Third World have been characterised by an excess of labour. Turnover and absenteeism are no longer of concern; in the major firms they are frequently at levels below those prevailing in industrialised countries. Sabot (1979), in a landmark study, describes and analyses the transformation in Tanzania. The colonial labour market was characterised by a persistent shortage of labour. Government and private employers paid low wages, arguing that at higher wages migrants would work for shorter periods. Sabot could find no evidence to support this proposition. Moreover, even if higher wages were to induce the individual migrant to work for a shorter period, they could also be expected to attract a greater number of migrants. It appears then that a small number of large estate owners and the government took advantage of their oligopolistic position to administer wages at a level below that which would attract a sufficient number of workers. The shortfall was made up in part through forced labour until the 1920s and again during World War II, and through unscrupulous methods of recruiting labour in distant regions. After World War II urban wages increased steadily while rural incomes stagnated. When independence was granted in 1961, the rise in wages accelerated. Growth of the rural labour force and the widening gap between income from farming and from urban wage employment increased the size of the migrant stream, and labour surplus replaced labour scarcity. At the same time a labour force that had been characterised as "uncommitted" was stabilised: turnover and mobility rates became low by international standards. Sabot explains the dramatic rise in wages and the transformation of the problem of labour scarcity into a problem of urban unemployment in terms of the convergent interests of employers, trade unions, and government. The advent of independence prompted a burst of import-substituting industrialisation; these relatively capital-intensive factories had to invest in training the industrially disciplined semi-skilled workers they required; in order to secure this investment, the labour force had to be stabilised; and employers raised wages to that end. The trade unions were identified with the nationalist movement and gained in strength with the emergence of educated leaders and the stabilisation of the labour force; they had only begun to develop in the 1950s, but more than half the labour force was unionised by 1965, and they could bring pressure to bear on wages by both political action and collective bargaining. The government legislated a minimum wage in 1957 and increased the wages of low level government employees.

MIGRATION STRATEGIES

Rural-urban migration continues unabated throughout much of the Third World. Why do so many come, the question usually goes, when urban unemployment is widespread and underemployment common? To which a peasant might respond with the counter question: why do so few go, when the rural-urban gap is unmistakable? Two interpretations can be advanced to explain migratory behaviour under these circumstances; both establish that the decision to migrate is a rational response to economic conditions. The difference between the two interpretations is accounted for by variations in the structure of urban labour markets.

In Tropical Africa analysis focused on migrants coming in search of jobs that offered wages and working conditions regulated by legislation and/or collective bargaining. They would spend several months trying to secure such a job, but, if unsuccessful, eventually return to the village. Thus in Kampala, Uganda, Hutton (1973:61-2) found a clearly established pattern in the middle 1960s. Of the unemployed men she interviewed, three-quarters planned to leave if they could not find work, typically within less than six months. More than three-quarters of these intended to return to their rural home. Going home, however, was only a temporary measure; only 11 per cent of the unemployed surveyed felt that they would stay there.

In the 1950s and 1960s much urban unemployment in Tropical Africa appears to have conformed to this pattern. With independence, urban wages rose substantially in many countries - frequently the specific causes for wage increases were similar to those detected by Sabot in Tanzania, invariably the advent of independence raised expectations that governments judged had to be met at least to some extent. Rural-urban migration surged, the labour shortages that had plagued colonial governments vanished, and urban unemployment appeared. Since much labour migration had been short-term, recent immigrants faced little competition from entrenched workers and their descendants. Furthermore, independence was frequently accompanied by a significant expansion in urban employment. The system of recruiting unskilled labour approximated a random process. Since minimum wages were high, relative to rural incomes, even an extended job search was a promising strategy. Joining the urban unemployed, the rural-urban migrant tried his luck at the urban job lottery (Gugler 1969).

The proposition that potential migrants take into account the probability of securing urban employment, along with the rural-urban real-income differential, was incorporated into a model by Harris and Todaro (1968;1970).[5] The probability of obtaining urban employment was defined as the proportion of the urban labour force actually employed. The assumptions underlying this definition were problematic even for the early stages of urban unemployment in Tropical Africa (Gugler 1976). The most comprehensive test of the basic proposition was based on a survey of 5,500 households in seven Tanzanian towns in 1971. An analysis of the propensity to migrate according to level of education provided evidence for the significant role of both, rural-urban income differentials and urban employment probabilities. In contrast to Harris and Todaro, however, the employment probability was defined as the ratio of the net number of jobs created over a four-month period (the estimated average time spent in the job search) to the number of unemployed (Sabot 1979:120-27).

In retrospect it is clear that the urban job lottery pattern occurred in exceptional circumstances. More commonly, labour turnover is low, job creation slow, and recruitment anything but random in Third World countries. A more generally applicable interpretation of rural-urban migration has to focus on the fact that labour markets, like most markets, are fragmented in a variety of ways, i.e. different categories of people enjoy differential access to earning opportunities. Access is usually largely a function of three criteria: education and training, gender, and patronage. Differential access in turn shapes the composition of the migrant stream. The role of formal education as a prerequisite for access to the more privileged strata motivates parents in rural areas and small towns to relocate with their children or to send their children away to better and more prestigious schools. For those who have ascended the educational ladder, the most attractive career opportunities are

[5]For an account of subsequent modifications of the model, see Todaro (1976:36-45). Recently Harris and Sabot (1982) have proposed a generalised model of migration and job search in the context of wage dispersion and imperfect information, of which the Harris-Todaro model is a special case.

in the city. Where new industries have recruited substantial numbers of women, as in South Korea and in Thailand, the proportion of women among rural-urban migrants has increased.

'Credentials' are generally accepted as a screening device, and discrimination on the basis of sex is commonly taken for granted. Patronage, in contrast, is usually frowned upon. It is, however, widespread, sustained as it is by strong interests and effective mechanisms. Most migrants, as we noted before, expect and obtain assistance from urban contacts. The urban host, to help the new arrival, to relieve the burden of housing and perhaps even of feeding him, has good reason to find him work. Thus migrants who have secured employment introduce their relatives and other people from 'home' to their firm. Many employers find such 'family brokerage' convenient and even advantageous. They know that skills and knowledge are not as important for many positions as other qualities: dependability, potential for training, persistence, initiative. Further, in many cases, job advertisements will generate all too many applications from people with similar qualifications. In such circumstances the employer prefers to use a broker. He selects among his employees one or two persons he trusts, and asks them for suitable candidates, whom they will probably have to train. The broker will look to his extended family for suitable candidates, and draw up a short list. He may coach a candidate on how to fill in the application forms, and on how to react at the interview. A close and complex relationship thus arises between the employer, the broker, and the new employee. The broker has increased the socio-economic position of his kin group and his own standing within it, the unemployed has obtained a job, and the employer can exert leverage over his employee through the broker (International Labour Office 1972: 509-10). Because of such particularistic recruitment patterns, migrants of common origin tend to cluster in certain jobs and trades, the labour market is segmented.

The segmentation of the urban labour market is mirrored in the stream of migrants. Their prospects vary widely, but, at least, they are reasonably well defined for many. The integration of new arrivals into the urban labour market is thereby eased. Indeed, unemployment is frequently reported lower among the migrants than among the urban-born. Discontent over discrimination does not crystallize as long as the criteria for privileged access vary from one little niche in the urban economy to the next. In many countries, however, large segments of the labour market appear as the exclusive preserve of a racial or ethnic group, and bitter conflicts among immigrants and locals, or among immigrant groups, ensue.

WHO MIGRATES?

Given the rural-urban gap, wholesale emigration from the disadvantaged rural areas might be expected. But the cities are less than hospitable to new immigrants; only the highly trained, well connected, and hardy venture there. The stream that appears enormous at the urban end constitutes only a small proportion of the rural population.

There are major differences among migrants in terms of socio-economic background, and accordingly their urban prospects vary. At one end of the spectrum are the many who are poor, and ill-equipped for any but the most menial tasks: some come from neglected and impoverished regions, e.g. much of Burkina (Upper Volta), others originate from the lower strata of quite differentiated communities, e.g. village India; with few exceptions they have little schooling and are barred from most of the more rewarding opportunities. At the other end of the spectrum is the migrant from an unusually developed region, or more typically a member of a privileged rural minority, who attends the better schools and climbs the educational ladder high enough to gain access to a promising career in public administration, with a major company, or as a professional.

Young adults always predominate among migrants in search of employment. They are usually unmarried; but even when married, they have less at stake in the rural areas than their elders. They frequently lack control over resources, land in particular, and wield little power in local affairs. To put it into universal terms: they are at a transitional stage between adolescence and adulthood and not yet firmly committed to an adult role in the local setting. For that very reason they enjoy an advantage in the urban economy: less established, they are more adaptable to the different demands of the urban environment. And if migration entails accepting marginal earnings in the hope of eventually securing a protected job or satisfactory self-employment, then the potential rewards are highest for the young starting on a lifetime urban career.

The migration of women has received scant attention in the huge migration literature. When marriage requires migration, it is usually the woman who moves. For that reason in a country such as India women outnumber men in total rural-urban migration, even though men predominate in net rural-urban migration. If we are concerned with the latter, with the redistribution of the population through migration, urban sex ratios suggest considerable variation in the sex selectivity of net rural-urban migration (Table 7.1).

South Asia conforms best to the stereotype that women move usually as dependents, and are therefore outnumbered by men some of whom are unmarried or have left their wives behind. In this region, men outnumber women in the cities by a large margin in every country. The imbalance is somewhat reduced, but remains substantial, when urban sex ratios are adjusted for the fact that men outnumber women also in the total population of each of these countries (Ferree and Gugler 1985). In most countries in the Middle East and North Africa, men also outnumber women in the urban population, but, except for Turkey, by a smaller margin; this margin is further reduced when urban sex ratios are adjusted for national sex ratios. Morocco is a notable exception, presumably because of the absence of urban men who are working in Europe. The major countries of Subsaharan Africa report a considerable excess of men over women in their cities. Ethiopia, however, constitutes an important exception that may reflect a feudal pattern in which affluent urban households attract a disproportionate number of women in one guise or another. In East Asia, the low urban sex ratio in the Philippines is remarkable.

Latin America and the Caribbean stand in sharp contrast to other Third World regions. They resemble developed countries in that women outnumber men in the urban areas, throughout the region. In some countries women predominate by a large margin. It would appear then that substantial numbers of women move on their own, a pattern rarely encountered elsewhere in the Third World. The uniformity of the pattern, and its occurrence in the Philippines, the only Latin Third World country outside the Americas, suggest a cultural interpretation. While most Third World cultures encourage early marriage and child-bearing, the common heritage of Latin countries includes a religious ethos that exalts the status of the single woman.[6] Young Latin women thus are potentially mobile independent of a spouse. Faced with limited rural opportunities, they turn to the cities where many households can afford to offer them the low pay and limited benefits that go with domestic service. Everywhere young adults are the most likely to move because they are at a transitional stage between adolescence and adulthood, but the length of this stage varies a good deal across cultures - especially for women.

PATTERNS OF MIGRATION

The movement of individuals is the focus of much migration analysis. This tendency is encouraged by the fact that migration frequently involves young single persons. However, in many cases migration is not just a once for all move, rather there are a number of moves over a lifetime. Such a

[6] I am indebted to Francine van de Walle for this observation.

migratory career is best understood with reference to family and community. Three principal patterns of rural-urban migration in the Third World then stand out: (a) temporary migration of men; (b) family migration to urban areas followed by return migration to the community of origin; and (c) permanent settlement.

These are not fixed statuses. The man who left his family behind may decide to have them join him; the family that expected to return to its community of origin may settle down in the city forever. While changes in migratory status, as perceived by the migrant, typically go in the direction of an increasing commitment to the place of destination, they are clearly affected by changing circumstances in both the urban environment and the area of origin, e.g. deteriorating urban conditions may force men to send their families back to the village.

The preponderance of men over women in net rural-urban migration in South Asia, in much of Subsaharan Africa, and in Oceania (Ferree and Gugler 1985) reflects the tendency of male migrants to leave wife and children in their rural area of origin. If the Industrial Revolution engendered the distinction of workplace and home, the separation of worker and dependents has been drastically magnified for many Third World families. Extended family support typically facilitates such simultaneous involvement in the urban and the rural economy. Indeed, the assistance of male kin in certain tasks, and the protection they afford, frequently appears as a prerequisite for a wife to manage the farm and to hold her own in a male-dominated environment.

The migration of individuals, whether single or separated from their family, has distinct economic advantages: it optimises labour allocation, and, at least in rural-urban migration, it minimizes the cost of subsistence. Employers save on wages and retirement benefits, and public authorities face less demand for housing and infrastructure. But there are also gains to migrants that motivate them to accept family separation. Living costs in the city are high, while urban earning opportunities for women are usually very limited in these regions. Typically wife and children remain on a family farm growing their own food, and perhaps even raising cash crops. Where land is communally controlled and cannot be alienated, as is the case in much of Tropical Africa, there is no compensation for those who give up farming it. A wife who comes to town has to abandon an assured source of income to join a husband on low wages.

Throughout Latin America the predominance of women in the urban population suggests that the temporary migration of men to cities is not a common pattern. However, some Indian communities present an exception to this generalisation. They demonstrate the crucial role of rural social structure in the establishment and maintenance of specific patterns of migration. Laite (1981), in his study of labour migration in highland Peru, describes and analyses a wide-spread pattern of temporary migration of men. Thus, the majority of miners and refinery workers are migrants, and nearly all these migrants maintain village interests. The most important interest is in village land, even though land has increasingly been transformed into a commodity and brought into the cash arena since the 19th century. Three-generation extended families continue to control property and to organise production. The senior generation owns the resources while other members of the family work on them. So, whilst junior members of the household are 'landless', they do have access to land. The migration of one, or several, men provides external resources to meet household needs. During their absence, it is the women who do the work, or, at planting and harvest time, recruit labour. Their task is facilitated by co-operative practices well established in Andean peasant culture.

Family separation has commonly taken the form of circular migration.[7] After a period of employment lasting six months, perhaps, or a couple of years, the migrant returns for an extended stay with his family. In the ideal case, the return coincides with peak labour requirements on the farm. In some areas, such migrants go as contract labour, i.e., they are recruited for a fixed period of time at, or close by, their home place, and provided with return transportation. Repetition of the circular movement is common, and many migrants build up extended urban experience. It is tempting to speculate that circular migration is the initial response of a 'traditional' society to new opportunities to earn wages and acquire manufactured goods, to visualise 'tribesmen' making short forays into an alien environment. The facts indicate otherwise. Migration has been a permanent feature of highland Peru since peasants went to work in the mines during the Inca period; half a millenium later a pattern of temporary migration persists. In Indonesia, where circular migration was well established during the colonial period, it has since greatly increased in importance (Hugo 1983). And a review of migration research in the Pacific islands concludes by emphasising the persistence of circular migration during 150 years of European contact (Bedford 1973:126).[8] Circular migration in these areas today, as in Tropical Africa in the past, is a function of the recruitment of men at low wages.

---

[7]For a recent review of some of the theoretical debates over circular migration see Mitchell (1985).

[8]Goldstein (1978) reviews the literature on various forms of temporary migration in Asia, and in particular in South-East Asia. He emphasises that return migration, and in particular circular migration is much more common than the standard studies based on census data reveal.

Where employees have access only to bachelor accommodation, both aspects - the cheap labour policy and the limitation of recruitment to men - are brought into sharp relief. Most strikingly, in environments thus characterised by circular migration, major employers are able to establish a more stable labour force by providing conditions that encourage workers to bring their wives and children.

Circular migration was common in Tropical Africa in colonial days. However, the Union Minière du Haut Katanga changed its labour policy in the copper mines in what is now Shaba Province, Zaire, as early as 1927. A measure of compulsion was involved in that workers had to bring their wives, but the region soon boasted a stable labour force. By 1957 the average length of service of African employees in the Katanga mines was eleven years. In the Copperbelt, in what is now Zambia, a policy to establish a permanent force was initated in 1940: permanent accommodations for married employees were made available and the standard of housing improved, adequate schooling in the urban areas was provided, and a pension scheme instituted. The average length of employment of African workers increased from four and a half years in 1956 to seven in 1964. In sharp contrast, the South African gold mines continued to recruit labour on short-term contracts from adjacent areas throughout these years (Wilson 1972:123-27).

In parts of British Colonial Africa, as well as on some Pacific islands, circular migration was not just the outcome of policies adopted by employers: temporary sojourn of indigenous workers in urban areas was politically intended, and policies were adopted to that purpose. In South Africa, the gold mines are prohibited by law from providing family accommodation for more than 3 per cent of their African work force. Many men are recruited on short-term contracts, not only in the mines, but in various other sectors of the economy. The number of African migrants in the country has been estimated at about half of those in registered employment (Turok and Maxey 1976:241). Furthermore, over the last two decades, major efforts have been directed toward uprooting the many Africans who have been long settled in urban areas. Racial oppression has thus created a paradox: the most industrialised country on the continent, where large numbers of Africans have worked in mines, factories, and services for several generations, has the highest proportion of short-term recruits in its labour force.

Circular migration constitutes an adaptation to family separation: the migrant returns regularly to his wife and children for extended periods of time, and he remains actively involved in the extended family, and indeed in village affairs. This strategy fails with the appearance of urban unemployment. When the search for a job takes months, circular migration is no longer a viable proposition. The migrant who wants to be assured of urban employment has to cling to his job. Instead of extended stays with the family there are short visits, as employment conditions and distance permit. What had been an economic cost to employers - a labour force characterised by high turnover and absenteeism - has become an increase in social costs for workers: more severe strains in their relationships with wife, children, extended family, and village community. The frequency of visits varies a great deal. As improved means of transport are introduced, and their cost decreases, monthly or even weekly commuting become more common. But in a country such as India, many migrants cover considerable distances and can visit their families only during their annual leave.

The preponderance of men in the urban population has sharply declined in a number of countries over the last two decades, suggesting that family separation is becoming a less common pattern. There is no evidence to suggest that urban earning opportunities for women have markedly improved, save in a few rapidly industrialising countries such as South Korea and Thailand. Certainly, as the period of urban employment has lengthened, family separation has become a less satisfactory pattern. At the same time, increases in urban wages have diminished the significance of the rural income forgone and of the urban/ rural cost of living differential. Also, in some areas, land shortages entailing a decline in output or the, often related, breakdown of communal control over land transforming it into an asset that can be realised, have reduced the opportunity cost of abandoning farming.

Settling down in town with a family is usually for the long term, many times for a working life, but it does not always signify a permanent move. Strong ties to members of the extended family who have stayed behind, and to the village community, can make an eventual return an attractive proposition. At the same time most migrants, even when they manage to support a family in the urban setting, enjoy little economic security. Unemployment and underemployment are widespread, but few qualify for unemployment compensation. And social security systems covering disablement and old age are still in their infancy. For many urban dwellers, the solidarity of the village provides an alternative social security, meagre but reliable. Plentiful land under communal control is still a common pattern in Tropical Africa and the Pacific. In such a situation the migrant can maintain his position in the rural community, and even during an extended urban career remain assured of access to land on his return. A review of survey data for five West African countries showed in every case a substantial proportion of the migrants indicating the intent to retire in their home area (Gugler and Flanagan 1978b). In communities in which return migration is a common pattern, there is a tendency for it to be established as a cultural norm, just as in the case of rural-urban migration. Rather than acknowledging the economic imperative, the norm of return migration is typically articulated as an ideology of loyalty to the home community.

In parts of Africa, in much of Asia, and especially in Latin America, most migrants have little prospect of maintaining access to agricultural land because of population pressure and/or institutional constraints. Wholely dependent on their urban earnings, they have become proletarianized. Instead of planning for a return to the village, they press for the provision of social security to urban workers. And they search for sources of earnings outside of employment. Escaping the vagaries of employment is a major attraction in establishing one's own business. Ownership of a home, however rudimentary, gives the assurance of accommodation and offers the possibility of deriving income from rent. The strength of squatter movements in Latin America may be understood at least in part in terms of the need for migrants who have severed their rural connection to establish an urban base.

The demands made by migrants on the urban system thus will vary according to their plans for the future. Single migrants expect little. Indeed, the limited prospects they face are the key reason the married left their family behind. They will tend to opt for a minimum of expenditure for housing, which will allow target migrants to shorten their stay and enable others to increase their remittances and savings. Migrants who anticipate eventually returning to their area of origin will remain concerned with conditions there. For permanent migrants the provision of security in the urban setting, especially in old age, becomes crucial.

In delineating these three patterns of migration I have given considerable emphasis to intended, as against actual, migratory behaviour. There is some question, however, whether a large proportion of migrants realise their plans to return to their area of origin. Indeed, it is frequently assumed that the majority will end their lives in the cities (Lloyd 1979:136). I am not so sure. What little data there are suggest that many migrants do retire to the rural community. In any case, many of the implications of return, as distinct from permanent, migration hold, whether the intention to return is ultimately realised or not; they hold as long as migrants act upon their assumption that one day they will settle down 'back home'.

MIGRATION POLICY

The pace of urban growth in the Third World, without precedent in human history, is a matter for concern. Policies to slow down natural population growth are an obvious response, since it accounts for more than half of urban growth in most Third World countries (United Nations 1980:24). However, unless coercion is employed, changes in procreative behaviour are rather slow, and they affect the labour market only after a considerable lag. In contrast, the majority of rural-urban migrants are ready to enter the labour force. If they contribute between one-third and one-half to urban growth, they constitute a considerably higher proportion of the new entrants into the labour force. Not surprisingly then most Third World governments grope for policies to slow down rural-urban migration. A 1983 United Nations survey of 129 governments of less developed countries found that only six considered the distribution of their populations 'appropriate'; three-quarters stated that they were pursuing policies to slow down or even reverse internal migration (United Nations 1983: table xxii.3).

The case for policies to reduce rural-urban migration is strengthened if the rate of urbanisation in the Third World today, while not without historical precedent, is seen as constituting overurbanisation: rural-urban migration, which involves the opportunity cost of rural output forgone, brings workers to cities that are unable to fully employ their existing labour force to productive ends; furthermore these additional urban dwellers require more resources for their survival than they would in the countryside (Gugler 1982). Migrants, however, rationally maximize their benefits. The resolution of the seeming paradox derives from the fact that rural-urban migration has a redistributive effect. Rural-urban migrants lay claim to a share in urban income opportunities, they gain some access to urban amenities. Rural families send their sons and daughters to the city so that they will be able to partake, however little, of its riches.

Policies to restrict rural-urban migration constitute an attempt to keep the have-nots out, to erect a boundary that will shelter urban populations from the competition of migrants, just as affluent countries shelter their citizens by the enactment and enforcement of laws controlling the immigration of foreigners. Boundaries are drawn and barriers erected to protect privilege. Restricting rural-urban migration means closing the remaining escape route to the rural masses that have been disenfranchised and neglected. Still, if migration controls are the exception in the Third World, it is not out of concern for equity. Rather, internal boundaries, to a much greater extent than international boundaries, are difficult to police. The countries that regulate internal migration are those that control their populations rather closely; they include South Africa, committed to maintaining inequity between Black and White, as well as socialist countries that avow their concern with the 'rural-urban contradiction'. Even where drastic measures are taken to enforce controls on internal migration, however, their effectiveness appears limited. Half of the residents in the African township of Soweto, on the outskirts of Johannesburg, are said to live illegally there. In China hundreds of thousands of rusticated middle-school leavers returned to the cities without authorisation in the 1970s.

Contract labour in China, the temporary recruitment of workers from rural areas to work in urban jobs, illustrates the dilemmas of equity and efficiency that migration control poses. Blecher (1983; 1986) provides an account for Shulu, a county in Hebei Province. Between 1964 and 1978 the industrial labour force in Shulu trebled. Three quarters of the new workers came from rural areas on temporary contracts. By 1978, contract workers comprised over half of the work force in county-level industry. They received somewhat lower wages than regular workers. Moreover, they had no claim to the fringe benefits enjoyed by regular workers, such as free medical insurance, accident insurance (workers' compensation), pensions, and sick leave. Most importantly, contract workers were housed in dormitories and had to leave their families in the rural areas. In a period of rapid industrialisation urban growth was thus reined in. Only workers needed in production were authorised to come to urban areas, and most, on contract terms, had to leave their dependents behind. To the extent that the system was effectively implemented, unemployment was avoided and the proportion of dependents kept extremely low in the urban areas. At the same time, inequities were created between regular workers and contract workers, as well as between the latter and rural workers. Rural production teams, however, received a share of contract workers' wages; and in as much as contract workers and their families remained part of the rural population, the average income of that population was higher than it would have been if they had left altogether. The system can be argued to be more beneficial to the rural masses than if some in their midst were to leave permanently and cut all ties.

There is general agreement that rural development, so as to close the rural-urban gap, constitutes a policy of equity that will truly stem rural-urban migration. Admittedly, initial improvements in education and in communications may encourage migration as long as rural-urban discrepancies remain large. To significantly reduce the gap requires that considerable resources be made available to a huge rural population. Rural incomes have to be raised directly, e.g. through a reduction in taxes or through an increase in the prices of agricultural products, and/or indirectly through investments that raise the productivity of the rural labour force whether in agriculture or in small-scale industry. Given the size of the rural population in Third World countries, this would require an enormous reallocation of resources, a reallocation that would be confronted with the determined opposition of urban interests. A long-run rural development policy, which in any case would have to focus on labour-intensive investments to retain a growing rural population, might constitute a feasible compromise. Such a policy appears, however, to be beyond the horizon of most governments, preoccupied as they are with their very survival in a much more immediate future.

REFERENCES

Bedford, R.D. (1973) New Hebridean Mobility: A Study of Circular Migration, Department of Human Geography Publication HG/9, Research School of Pacific Studies, Australian National University, Canberra

Berg, E.J. (1961) 'Backward-Sloping Labor Supply Functions in Dual Economies - The Africa Case', Quarterly Journal of Economics, 75, 468-92

Blecher, M. (1983) 'Peasant Labour for Urban Industry: Temporary Contract Labour, Urban-Rural Balance and Class Relations in a Chinese County', World Development, 11, 731-45

Blecher, M. (1986) 'Rural Contract Labor in Urban Chinese Industry: Migration Control, Urban-Rural Balance and Class Relations' in J. Gugler (ed.) Perspectives on Third World Urbanization, Oxford University Press, Oxford

Ferree, M. Marx, and J. Gugler, (1985) 'Sex Differentials in Rural-Urban Migration: Variations Across the Third World', Paper presented at the South South Conference, Montreal, May 1985

Gilbert, A. and J. Gugler, (1982) Cities, Poverty, and Development: Urbanization in the Third World, Oxford University Press, Oxford and New York

Goldstein, S. (1978) Circulation in the Context of Total Mobility in Southeast Asia, Papers of the East-West Population Institute 53, East-West Centre, Honolulu

Gugler, J. (1969) 'On the Theory of Rural-Urban Migration: The Case of Subsaharan Africa' in J. A. Jackson (ed.) Migration, Cambridge University Press, Cambridge, pp. 134-55

Gugler, J. (1976) 'Migrating to Urban Centres of Unemployment in Tropical Africa' in A. H. Richmond and D. Kubat (eds.) Internal Migration: The New and the Third World, Sage Studies in International Sociology 4, Sage Publications, London and Beverly Hill, CA, pp. 184-204

Gugler, J. (1982) 'Overurbanization Reconsidered', Economic Development and Cultural Change, 31, 173-89

Gugler, J. and Flanagan, W. G., (1978a), Urbanization and Social Change in West Africa, Cambridge University Press, Cambridge, London, New York, Melbourne

Gugler, J., and W. G. Flanagan (1978b) 'Urban-Rural Ties in West Africa: Extent, Interpretation, Prospects, and Implications', African Perspectives 1, 67-78

Harris, J.R., and R.H. Sabot (1982) 'Urban Unemployment in LDCs: Towards A More General Search Model' in R.H. Sabot (ed.) Migration and the Labor Market in Developing Countries, Westview Press, Boulder, Colorado, pp. 65-89

Harris, J.R., and M.P. Todaro (1968) 'Urban Unemployment in East Africa: An Economic Analysis of Policy Alternatives', East African Economic Review, 4, 17-36

Harris, J.R., and M.P. Todaro (1970) 'Migration, Unemployment and Development: A Two-Sector Analysis', American Economic Review, 60, 126-42

Hugo, G.J. (1983)'New Conceptual Approaches to Migration in the Context of Urbanization: A Discussion Based on the Indonesian Experience' in P.A. Morrison (ed.) Population Movements: Their Forms and Functions in Urbanization and Development, Ordina Editions, Liège, pp.69-113

Hutton, C. (1973) Reluctant Farmers? A Study of Unemployment and Planned Rural Development in Uganda, East African Studies 33, East African Publishing House, Nairobi

International Labour Office (1972) Employment, Incomes and Equality: A Strategy for Increasing Productive Employment in Kenya, ILO, Geneva

Laite, J. (1981) Industrial Development and Migrant Labour in Latin America, Manchester University Press, Manchester, and University of Texas Press, Austin, Texas

Lipton, M. (1977) Why Poor People Stay Poor: A Study of Urban Bias in World Development, Temple Smith, London, and Harvard University Press, Cambridge, MA

Lipton, M. (1984) 'Urban Bias Revisited', Journal of Development Studies 20(3), 139-66

Lloyd, P. (1979), Slums of Hope? Shanty Towns of the Third World, Penguin, Harmondsworth

Lowder, S. (1978) 'The Context of Latin American Labor Migration: A Review of the Literature Post-1970', Sage Race Relations Abstracts, 6, 1-49

Mitchell, J.C. (1985) 'Towards a Situational Sociology of Wage-Labour Circulation' in R.M. Prothero and M. Chapman (eds.), Circulation in Third World Countries, Routledge & Kegan Paul, London, Boston, Melbourne, Henley, pp. 30-53

Nolan, P., and G. White (1984), 'Urban Bias, Rural Bias or State Bias? Urban-Rural Relations in Post-Revolutionary China' Journal of Development Studies 24(3), 52-81

Nyerere, J.K. ([1967] 1968) 'The Arusha Declaration' in J.K. Nyerere (ed.) Ujamaa - Essays on Socialism, Oxford University Press, London, Oxford, New York, pp.13-37.

First published in Swahili by TANU, the party organisation

Pearse, A. (1970) 'Urbanization and the Incorporation of the Peasant' in A.J. Field (ed.), City and Country in the Third World: Issues in the Modernization of Latin America, Schenkman, Cambridge, MA, pp. 201-12

Preston, S.H. (1979) 'Urban Growth in Developing Countries: A Demographic Reappraisal', Population and Development Review 5, 195-215. Reprinted 1986 in J. Gugler (ed.), Perspectives on Third World Urbanization, Oxford University Press, Oxford

Rouch, J. (1956) 'Migrations au Ghana (Gold Coast): Enquête 1953-1955', Journal de la Société des Africanistes 26, 33-196

Sabot, R.H. (1979) Economic Development and Urban Migration: Tanzania 1900-1971, Clarendon Press, Oxford

Shaw, R. P. (1975) Migration Theory and Fact: A Review and Bibliography of Current Literature, Bibliography Series 5, Regional Science Research Institute, Philadelphia

Shaw, R. P. (1976) Land Tenure and the Rural Exodus in Chile, Colombia, Costa Rica, and Peru, Latin American Monographs, 2nd Series 19, The University Presses of Florida, Gainesville

Simmons, A.B., S. Diaz-Briquets and A.A. Laquian (1977) Social Change and Internal Migration: A Review of Research Findings from Africa, Asia, and Latin America, International Development Research Centre, Ottawa

Sivard, R. L., (1985) World Military and Social Expenditures 1985, World Priorities, Washington, D.C.

Todaro, M.P. (1976) Internal Migration in Developing Countries: A Review of Theory, Evidence, Methodology and Research Priorities, International Labour Office, Geneva

Turok, B., and K. Maxey (1976) 'Southern Africa: White Power in Crisis' in P.C.W. Gutkind and I.Wallerstein (eds.), The Political Economy of Contemporary Africa, Sage Series on African Modernisation and Development 1, Sage Publications, Beverly Hills, CA, and London, pp.232-60

United Nations (1980) Patterns of Urban and Rural Population Growth, Population Studies 68, United Nations, New York

United Nations (1983) 'World Population Trends and Policies: 1983 Monitoring Report' Document IESA/P/WP.82/Add. 1, United Nations, New York

Wilson, F. (1972) Labour in the South African Gold Mines, 1911-1969, African Studies 6, Cambridge University Press, Cambridge

World Bank (1981) China: Socialist Economic Development, the main report, World Bank East Asia and Pacific Regional Office, Bangkok

World Bank (1985) World Development Report 1985, Oxford University Press, New York

Yap, L.Y.L. (1977) 'The Attraction of Cities: A Review of the Migration Literature', Journal of Development Economics, 4, 239-64

Chapter Eight

COUNTERURBANISATION

A.J. Fielding

INTRODUCTION

Three unanticipated changes in the character of internal migration in 'western' developed countries have taken place during the recent period. The first is the downturn in inter-labour market area mobility rates. These had been increasing up until about 1970 and were expected to increase thereafter. Indeed it was at about the time that the downturn occurred that Zelinski published his influential paper on the 'mobility transition'; this paper very persuasively situated internal migration within its broader social context, and argued that western developed countries were entering a phase in which unidirectional rural-to-urban migration streams would be replaced by 'circulation', a situation in which high levels of mobility would characterise all areas. It was not to be. The levels of interregional migration were in almost all cases lower in the late 1970s and early 1980s than they were in the 1960s (Ogilvie, 1979).

The second unexpected change is the 'turnround' in the pattern of net migration. Prior to about 1970 there was a positive correlation between the population size of a settlement (conceived here in labour market area terms) and its net migration; the smaller, more rural or more isolated a place was, the more likely it was to lose population through internal migration, and the larger, more urban or more central a place was, the more likely it was to gain through internal migration. It might have been expected that this pattern would have continued after 1970 because the employment shift away from agriculture and towards the services continued, as did the advantage of metropolitan areas over rural areas with respect to wage levels and levels of unemployment. Again this did not happen. The relationship between size of settlement and net migration became a negative one, the largest cities lost and the smaller towns and rural areas gained (Court, 1984).

The third unexpected change concerns the social composition of internal migration streams. The reader may possibly have already decided that the changes in levels of mobility and in patterns of net migration have something to do with the recession in the world economy. This recession is customarily thought to have begun with the 'oil crisis' of 1973-4 and to have intensified in the years 1979-81. Since the impact of recession is uneven as between regions and social classes, it might have been expected that those most adversely affected, that is, manual workers in manufacturing industries and young people entering the labour market in non-metropolitan and 'old industrial' regions, would have come to constitute an even larger proportion of the interregional migrants than they did in the 1950s and 60s. The signs are that this has not been the case. The migration streams seem to have been increasingly dominated by members of the 'new middle class' (Pohl and Soleilhavoup, 1982).

Thus we have three paradoxes:

(i) a decrease in internal migration just as we enter the high mobility 'post-industrial' era;

(ii) a replacement of urbanisation by counterurbanisation despite the location in the major cities of the favourable employment structures, higher than average wages and lower than average unemployment rates;

(iii) a situation in which those that need to migrate the most, that is, the low paid and those with low job security seem to constitute a decreasing proportion of migration streams, while those who need to migrate the least, that is, those in secure, well-paid jobs, constitute an increasing proportion of migration streams.

How can these paradoxical developments be explained? This paper focuses on the literature that relates to the second problem - the replacement of a trend towards a spatial concentration of population by a trend towards deconcentration - but it will be seen that some of the ideas developed recently to solve this problem can, in addition, shed a light on other issues in contemporary migration analysis.

The literature reviewed in this essay is bounded by the following criteria:

(i) the research must relate to population redistribution since 1950, and to the changes that occurred about 1970 and which resulted in counterurbanisation;

(ii) it must be concerned with population redistribution in 'western, developed', i.e., mature capitalist societies. This in practice means the countries of North America and Western Europe plus Japan and Australasia;

(iii) it must relate to changes at the inter-labour market area level; it will not be concerned with suburbanisation, though it is necessary to discuss the relationship between suburbanisation and counterurbanisation since the degree to which these two processes are to be seen as empirically and conceptually separate is controversial.

This brings us to the question of definitions. Most of the important concepts used in this essay will be defined as they are introduced, but the term 'counterurbanisation' is so central to the whole exercise that it is sensible to attempt a definition right at the start. 'Counterurbanisation' will be used in its narrow sense; it will refer to the tendency for larger places to lose population through migration, while smaller places gain. 'Larger' and 'smaller' are here defined in terms of resident population numbers and 'places' are defined in terms of labour market areas. Counterurbanisation is thus the opposite of 'urbanisation' where this term is used to mean 'a process of population concentration that implies a movement from a state of less concentration in space to a state of more concentration' (Tisdale quoted in Berry, 1976).

The relative population growths of large and small places is not, of course, just dependent upon migration; but social and spatial variations in fertility and mortality rates have become very much less important in the post World War II period, and the differences in natural increase that have occurred have been largely due to differential age-structures consequent upon age-selective migration (Robert, 1983). Thus the population changes experienced in particular places have been very largely due either to the direct effect of migration upon the numbers of people resident there, or to the indirect effects of migration upon the place's demographic structure. This explains the emphasis in this essay on population redistribution.

Even without the special interest in population redistribution occasioned by the recent changes in trends, this would be an opportune moment to take stock of what was happening to the population geographies of these countries. The information from the 1980-82 round of censuses is now largely available, and the massive task of analysis and interpretation is well under way. In addition, developments in the concepts of the analysis of urban and regional change have been rapid in recent years, and the time is ripe for them to be tested in the field of empirical population geography.

THE TURNROUND IN THE USA

Most of the many commentators on the migration 'turnround' in the USA give the impression that the change in trend was totally unforeseen. This is not so. Writing in the late 1950s, Vernon (1963) predicted that the New York metropolitan region would experience annual net in-migration of 114,000 persons per annum during the period 1955-65, dropping to 40,000 in the 1965-75 decade. 'In the following ten-year period, there will be net out-migration from the region - a truly spectacular result, if it can be believed, since it would reverse a historical trend of hundreds of year's standing'. Far-sighted as this prediction was, it was not an unqualified success; in the event the reversal came rather earlier than expected, and the relative decline in manufacturing employment upon which the prediction was largely based, turned out to be an absolute decline.

That American geographers and regional scientists were bemused by the turnround is less surprising when one remembers the number, and persuasiveness of the explanations that had been developed to account for the continued growth of the very large cities. Some emphasised the 'economic base', and showed how the absence of declining sectors such as agriculture, and the predominance of expanding sectors such as certain specialised services and high-growth industries, favoured the large metropolitan cities. Others stressed the importance of market size in the location of new investment, the role of agglomeration economies, and major cities as the centres of innovation, gaining from the benefits of 'initial advantage' and experiencing a 'virtuous circle' of growth. Despite all these reasons for the further growth in population and employment of the largest cities and the further decline of rural and peripheral areas, the post-1970 population redistribution pattern in the USA took the following forms (Hall, 1983):

(i) a negative relationship between growth and urban size;
(ii) a positive relationship between growth and distance from the city centre;
(iii) a net migration gain by non-metropolitan areas at the expense of metropolitan areas;
(iv) a regional shift in favour of the south and west at the expense of the north and east.

Figure 8.1 shows the shifts that have taken place in the net migration rates for metropolitan areas in different size categories between the 1960s and the early 1970s. It is unfortunate that the lowest two categories (metropolitan areas with populations less than 250,000, and non-metropolitan areas, i.e., areas lacking a population cluster of more than 50,000) are not subdivided, but the picture that emerges is clear enough; the larger places have moved from net migration gain to loss, and the smaller places from loss or small gain to rapid gain (Beale, 1977; Wardwell and Brown, 1980).

In addition to the shift towards a negative relationship between net migration and size of place, there was population redistribution within places. This favoured the areas more distant from the centre of the city over those that were nearer (suburbanisation). Data for the early 1970s shows, however, that the trend towards net migration gain in areas outside the metropolitan areas was not due to maldefinition of

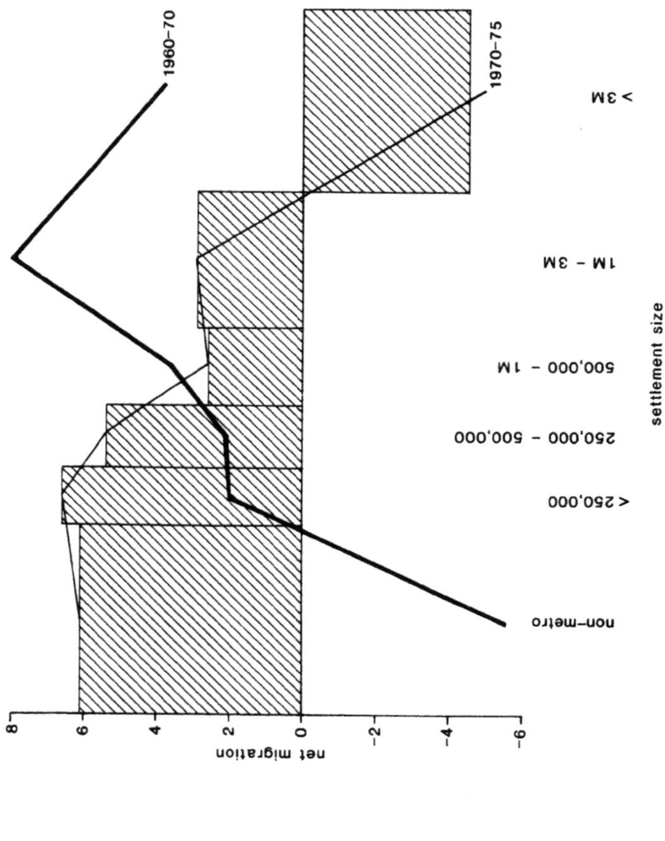

Fig. 8.1 US annual net migration rates per thousand population 1960–70 and 1970–75 by settlement size (SMSA) category.

the metropolitan area boundaries (statistical underbounding). From Table 8.1 it can be seen that although the levels of net migration gain favoured 'nearer' over more distant counties, the shift towards net gain was roughly equal over all counties irrespective of their commuting links with metropolitan areas.

This picture is confirmed by the 1980 census results. The population growth rates 1970-80 for counties not adjacent to metropolitan areas (+13.8%) was lower than that for adjacent counties but exceeded the rate for metropolitan counties as a whole (+10.0%).

The three major migration flows in the USA in the years 1900-70 had been the movement from east to west, the migration of black Americans from the southern states to the cities of the north, and the movement from the rural areas into the urban (Greenwood, 1981). Commenting on the 1980 census results, Hauser (1981) writes, 'at least one aspect of this well-established pattern was fundamentally altered in the 1970s. For the first time since the taking of the census began (1790), the population of non-metropolitan counties, made up of rural areas and small towns, grew faster than the population of the metropolitan areas'. The rapidity of this change is shown in Table 8.2.

It must be remembered, however, that as urban clusters attain 50,000 population they are transferred from the non-metropolitan category into the metropolitan. As a result the proportion of the total population of the USA that lived in metropolitan areas actually increased from 69% in 1970 when there were 243 metropolitan areas, to 75% in 1980 when there were 318 metropolitan areas.

Finally, and perhaps most dramatic of all, has been the regional shift in favour of the south and west at the expense of the north and east. In Figure 8.2 the five major regions of the USA have been ordered from the least urban (% non-metropolitan population) to the most urban. The 'sunbelt/frostbelt' contrast in the net migration figures is very marked (but see Plane, 1983).

The data in Figure 8.2 is once again consistent with data from the 1980 census which shows that the Northeast and Northcentral states had population growth rates 1970-80 that were far lower than the national average (+0.2% and +4.0% compared with +11.4%), while the South and West had rates that were far higher than the average (+20.0% and +23.9% respectively). The census results also show that the major changes in the 1970s compared with the 1960s were the sharp upward trend in the South (+5.7%) compared with the sharp downward trend in the Northeast (-9.6%).

Information for the early 1980s suggests that although the net migration gain by non-metropolitan areas at the expense of metropolitan areas may have ceased, the gains by the South and West at the expense of the urban-industrial Northeast and Northcentral regions of the country have

Table 8.1: Annual Net Migration Rates per Thousand Population for US Non-metropolitan Counties Classified by Degree of Functional Connection with Metropolitan Areas 1960-70 and 1970-75.

Level of Commuting to Metropolitan Area	1970 pop. x 10^6	Net migration rates 1960-70	1970-75	% per annum Shift
less than 3%	26.2	-8.3	+4.6	+12.9
3-9%	13.5	-4.2	+5.6	+9.8
10-19%	9.3	-1.3	+7.8	+9.1
20% and over	4.0	+1.1	+13.1	+12.0

Source: Berry (1980).

Table 8.2: Ratio of Metropolitan to Non-metropolitan Population Growth Rates in the US 1950-80.

	1950-60	1960-70	1970-80
Ratio metro/non-metro	3.7	2.4	0.7

Source: Long (1982).

persisted and may even have increased in importance. At the same time there are signs that the overall level of mobility has decreased (Plane, 1984).

The pervasiveness of population dispersal in the USA during the recent period is shown by Long and de Are's work on the 1980 census (Long, 1982). Using the Hoover index of concentration, they measured the spatial distributions of the population of the USA in 1970 and 1980 at different spatial levels (county, state, division and region), and found that deconcentration had occurred at each of these levels. It is hard to escape Berry's (1976) conclusion that around 1970 'a turning point (had) been reached in the American urban experience. Counterurbanisation (had) replaced urbanisation as the dominant force shaping the nation's settlement patterns' (Berry, 1976, 1980; Roseman, 1977; Vining and Strauss, 1977).

THE TURNROUND IN OTHER COUNTRIES

It soon became clear that the counterurbanisation phenomenon was not confined to the US but was general throughout 'western' industrial, mature capitalist countries (Vining, 1978,

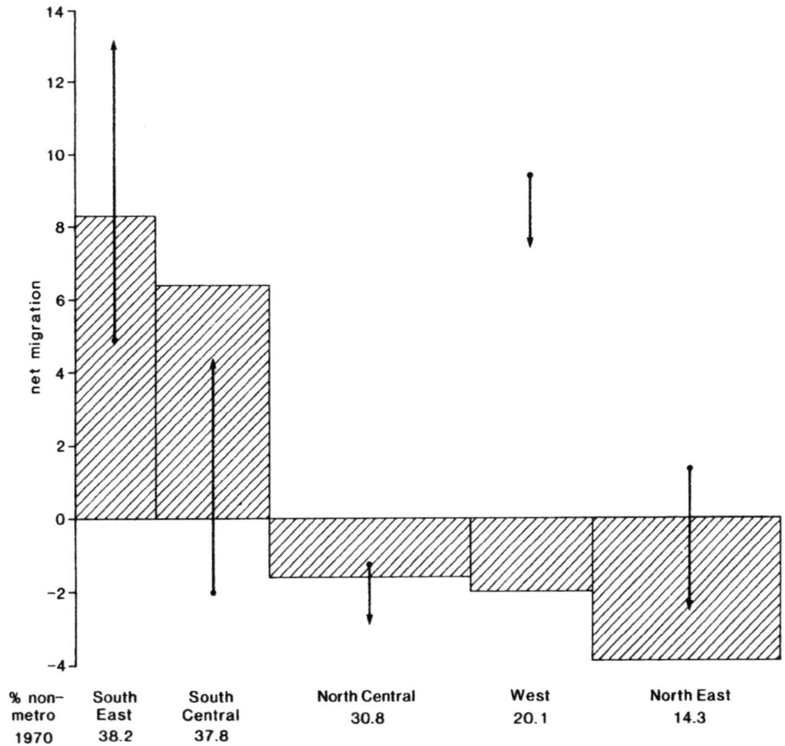

Fig. 8.2 Trend in US regional annual net migration rates per thousand population 1960-70 to 1970-75.

1982; ERIPLAN, 1978; Goldstein, 1975; Illeris, 1979, 1981). Although the concepts that were used varied somewhat, the substance of the empirical studies of trends in urban development was that urbanisation slowed down sharply during the 1960s, to be partially or completely replaced by counterurbanisation in the 1970s.

For a description of recent trends the countries can be grouped into three categories:

(i) those that have a relatively recent history of urbanisation, share a common western (European) cultural tradition, and contain large 'interiors' devoted to primary production (agriculture, forestry, mining): Australia, Canada, New Zealand;

(ii) a country with a long urban history but with recent 'western' social and economic forms fused with an eastern cultural tradition: Japan;

(iii) those that have a long history of urbanisation, share a western cultural tradition and lack large interiors: France, United Kingdom, Eire, Belgium, Netherlands, West Germany, Denmark, Norway, Finland, Switzerland, Austria, Greece, Italy, Spain and Portugal.[1]

It is not possible to treat in detail the trends in urban development in each of the countries listed above; instead Australia, Japan and France will be used to 'represent' their respective groups.

In the period between the second World War and 1970 the pattern of urban development in Australia was as simple as it was remarkable. It was remarkable for the degree of concentration of the nation's population in the five large 'port capitals' (Sydney, Melbourne, Brisbane, Adelaide and Perth), and was simple in that these cities were growing rapidly through migration, while interior and rural places were generally areas of net migration loss. Sydney's share of the population of New South Wales increased from an already high figure of 40% in 1947 to 60% in 1971, and Melbourne's share of Victoria's population increased during the same period from 53% to 70% (Whitelaw, 1984). Rapid employment growth in both manufacturing and the services, accompanied by further urbanisation was enhancing the already pronounced 'primacy' of the Australian state capitals.

During the 1970s Australian geographers considered it unlikely that counterurbanisation was taking place in their country (Whitelaw, 1984), but subsequent analysis of

[1]Other countries might have been included in the category western industrial, mature capitalist countries, but small size and/or the special circumstances of their present or recent political histories make them rather unsuitable for this essay (Hong Kong, South Africa, Israel, Argentina ...).

migration in the 1971-76 and 1976-81 inter-censal periods have produced a contrary conclusion (Jarvie, 1984; Hugo, 1984). The first signs of the changes in migration patterns came with the publication of the 1976 census. Contrasting the 1966-71 and the 1971-76 periods, and using 'overbounded' definitions of the capital city areas, Jarvie shows that there was a reversal in the net migration balances whereby Sydney and Melbourne, which had been net gainers in the late 1960s (+4 per thousand and +5 per thousand respectively) had become net losers by the early 1970s (-28 per thousand and -14 per thousand respectively) (Jarvie, 1984). At the same time large areas of non-metropolitan Australia in coastal, and non-coastal but non-interior regions had experienced the reverse transition, that is, from net loss to net gain. In addition the rates of net loss for many of the interior regions in the southeastern third of the country had diminished.

Small area statistics on internal migration from the 1981 census have not yet become available, but figures on net migration (internal and international) 1976-81 by size of settlement show a clear counterurbanisation form.[2]

The census data also shows, however, that the turnround in Australia was no simple reversal of the previous pattern. While the major metropolitan areas (other than Brisbane and Perth) ceased to gain appreciably by migration and probably lost by internal migration, many of the remote areas of the interior, both the moderately settled ones of the east, southeast south and southwest, and the sparsely settled ones of the centre and northwest were also areas of net migration loss. This means that the areas of net gain were two of the larger cities (Brisbane and Perth), many small and medium-sized towns in the more densely-settled peripheral parts of the country, some trading centres and 'resource' towns of the interior, and many rural areas along the east and northeast coasts.

Perhaps the most striking manifestation of the turnround in Australia is the altered pattern of interstate migration. In the 1950s and early 1960s the rapid growths of the two

---

[2]This statement is based on data presented in Hugo 1983 p. 243, which shows that for the period 1976-81 the net migration rates (internal and international) were +1.6 per thousand for cities over 1 million, +3.5 per thousand for cities between 0.5 million and 1 million.... +8.0 per thousand for towns between 10 and 25 thousand and +17.4 for settlements between 5 and 10 thousand. In this case, however, local administrative areas were used as a basis for the analysis and some of the variation is thought to be due to the statistical underbounding of the principal cities (Hugo 1983 p.42).

largest cities, Sydney and Melbourne, were fuelled by international migration for which these cities were the principal points of entry. But interstate migration reinforced this trend with each of these cities gaining 56,000 persons through the interstate migration of the Australian-born during the years 1954-66 (Burnley, 1980). The interstate net migration figures from the 1976 and 1981 censuses tell a very different story. New South Wales and Victoria saw their sizeable net gains through international migration largely dissipated through net interstate migration losses. The gains through interstate migration were experienced by Queensland, Western Australia and the Australian National Territory, with the ANT's gains largely confined to the early 1970s and Queensland's gains increasing rapidly during the recent period. Indeed, Queensland gained 31,500 persons through interstate migration in just one year 1980-81; 16,000 of this gain came from NSW and 10,000 from Victoria (Hugo, 1983)!

It can be seen that although there was no simple reversal of trends around 1970 in Australia, several of the features identified in the US version of counterurbanisation were present in the recent period. The most important of these are the net internal migration losses of the largest cities, Sydney and Melbourne, and the rapid emergence of the Australian version of the 'sunbelt' phenomenon represented by the increased net migration gain Queensland, experienced by both its capital city, Brisbane, and by many other places in the State. Hugo (1984, p.275) writes:

'During the 1970s Australia entered a new era of population change quite different from that of earlier post-war years. There have been major re-alignments in patterns of fertility, mortality and international migration... (In addition, the 1981 census results) show conclusively that there has been... a reversal of the long established trend toward increasing concentration of population in Australia's major metropolitan centres... Some significant deconcentration of both population and jobs has occurred away from the nation's major cities and toward certain smaller towns and cities in rural areas'.

Japanese urbanisation in the 1950s and 1960s was far greater than that experienced in the US or Australia. The four prefectures of the South Kanto region centering upon Tokyo gained population through internal migration at a rate of between 250 and 400 thousand persons per annum during the years 1950-70 (Kawashima, 1980)! The same period saw the population employed in the primary sector decline from 50% to 18%. Around 1970 the situation changed; rural depopulation did not cease altogether but rates of loss for rural prefectures fell to a very low level, and the high rates of net migration gain experienced by the belt of metropolitan cities extending from Tokyo in the northeast to Osaka in the southwest (the Tokaido 'megalopolis') were replaced by low rates of gain or by net migration loss. Figure 8.3 shows the changes for three groups of prefectures: the first is the greater Kanto region (ten prefectures) centered on Tokyo; the second group consists of the ten prefectures of the Kinki and Tokai regions which contain the major conurbations of Osaka and Nagoya; the third consists mainly of the rural prefectures together with prefectures containing important regional cities and medium-sized 'free-standing' towns. This graph shows both the extraordinary shift that occurred between 1970 and 1975, and the way that the capital city region has retained some of its net migration gain while the other urban industrial city regions have not.

Surveys of Japanese urbanisation trends published in English do not include net migration rates by settlement size, but an indication of the spread of population growth to rural and peripheral prefectures is given by the fact that between 1965 and 1970, 21 of the 47 prefectures lost population at a time when the total Japanese population grew by 5.5%, whereas between 1975 and 1980 only one of the 47 prefectures lost population (Tokyo itself) despite the fact that the total Japanese population grew at the slower rate of 4.5%. The latest data on population growth rates for the year 1982 (Statistical Yearbook, 1983) suggest that there are two kinds of area that are growing more rapidly than the national average (i) the areas close to but not necessarily within the commuting zones of the largest cities (for example, some of the prefectures of North Kanto and Kinki regions) and (ii) provincial city prefectures such as Myagi (Sendai) and Fukuoka (Witherick, 1983).

Summarising these trends in urbanisation and shifts in migration, Japanese population geographers frequently refer to the 'U-turn' that took place in the early years of the 1970s (Kuroda, 1975; Kawabe, 1980; Nanjo, 1982; Yamaguchi, 1984). By this they mean that a shift towards return migration that led to increased out-migration from the major metropolitan areas, plus a sudden drop in the rural exodus, combined to produce an abrupt end to urbanisation in Japan. Urbanisation had dominated Japanese population redistribution almost continuously since the Meiji restoration in 1868, when Japan took its first steps towards 'western'-style industrialisation and 'modernisation'; its passing was interpreted as a sign of social maturity - Japan had joined the ranks of the world's developed nations (Kawashima, 1980).

Urbanisation in France was, like that in Japan, very rapid in the early post-war period. Large flows of young people from rural areas and small towns into the major cities, particularly Paris, accompanied the major expansion of urban manufacturing industries and the growth of service employment (Fielding, 1966). In the period 1954-62 there was a strong positive relationship between the size of settlement and net migration (Figure 8.4); only in the case of the largest

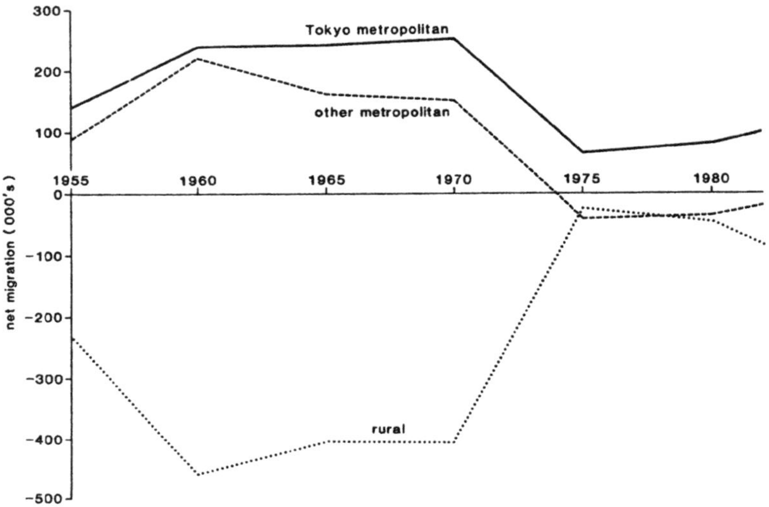

Fig. 8.3 Net migration by type of region Japan 1955-82.

settlements did the rates of gain not increased with size. The picture for the period 1975-82 is almost completely the reverse; over most of the range of settlement sizes the relationship is a negative one (that is, the counterurbanisation form) and it is only in the case of the very smallest rural communes that the rates of loss exceed those of the largest cities. Figure 8.4 also shows that it was between the mid-1960s and the early 1970s that most of the change took place (Courgeau, 1978, 1982; Boudoul and Faur, 1982).

The data on which Figure 8.4 is based ensures that settlements are defined in such a way as to include all the suburbs of a city when these consist of extensions to the continuously built-up area. In the recent period France has experienced 'periurbanisation', by which is meant urban development in areas that are near to, but do not form part of, the continuously built-up area. However, the data at the regional level for the two periods 1954-62 and 1975-82 shows that these local effects are not at the root of the changes that have taken place. In Table 8.3 France has been divided into just three broad zones: the Ile-de-France (Paris) region; the ten predominantly urban regions of northern and eastern France (containing the largest provincial cities such as Lyon, Marseille and Lille); and the predominantly rural regions of western France (these were in terms of employment structure the ten most agricultural regions in France in 1954). The change is quite striking. The net gain of the Paris region in

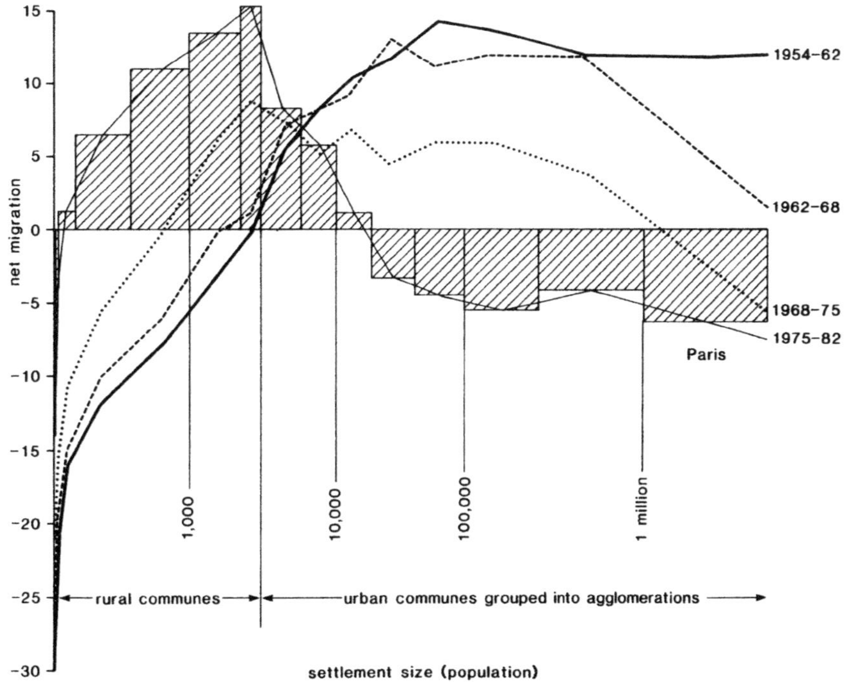

Fig. 8.4 France, annual net migration rates per thousand population 1954-82 by settlement size (rural commune and unites urbaines) category.

Table 8.3: Net Internal Migration Rates Percent Population for Major Regions in France 1954-62 and 1975-82.

Major Region	% pop. of France	Net Migration Rates 1954-62	1975-82
'rural' west	36.1	-2.26	+2.32
'urban' east	45.1	+0.11	+0.01
Paris region	18.8	+4.66	-4.51

Source: Censuses.

the late 1950s was turned into a loss, while the net loss of the rural western regions was turned into a gain (Francart, 1983).

One might argue that this threefold division was only appropriate to the situation in the pre-1970 period when the contrasts between Paris/province and urban east/rural west were at their greatest; since 1970 the principal difference seems to be between the net internal migration losses of the regions of northern France, notably Ile-de-France (Paris) (-440,000 1975-82), Nord-Pas-de-Calais (Lille) (-130,000) and Lorraine (-90,000), and the gains of the regions bordering the Mediterranean Sea, specifically Provence-Alpes-Cotes-d'Azur (+20,000) and Languedoc-Roussillon (+120,000).

In their first analysis of the 1982 census results the national statistical office write:

'The check to urbanisation is the most striking phenomenon revealed by the census results... During the period 1975-82, the growth of the rural communes is, on average, faster than that of the towns. Only towns of less than 20,000 inhabitants have a rate of growth higher than that for the whole of France. In over a century, this is the first time that such a phenomenon has been observed. It has its origins in the sharp reversal in the migration flows in favour of the rural communes...'.

Recent studies on other western countries confirm the trends discussed in this section. Broadly speaking, counterurbanisation replaced urbanisation around or just before 1970, continued until the late 1970s when either it disappeared or it was reduced to a lower level (Champion, 1983; Falk, 1978; Kontuly, 1984; Ogilvie, 1982).

THEORIES OF COUNTERURBANISATION

While many geographers and regional scientists have contented themselves with a documentation of the changes that took place around 1970, others have tried to provide reasons for these changes. The variety of positions adopted by writers on the turnround is immense; some see counterurbanisation as nothing more than the further development of already well-established trends notably suburbanisation, others see it as something completely new - as a 'clean break with the past'. For some the explanation is to be found in the factors influencing the location decisions of the individual migrants, notably the attractiveness of places with dry, sunny climates, fine scenery and open space in comparison with places that are regarded as drab, cold and overcrowded. For others the explanation lies in the inner workings of the mature capitalist economy, and it is the 'logic of capital accumulation' that drives the space-economy and results in population redistribution. Some researchers have been highly eclectic in their approach to explanation, while others have heroically defended 'single-cause' explanations such as economic recession or technological change. Similarly some insist that there are country- or even region-specific causes of the turnround while others point out the remarkable similarities in both the content and the timing of the changes that took place and assume that because of this the causes must lie in developments that are common to all western industrial societies. With so many approaches and such a rapidly expanding literature, it is not feasible to examine each writer's position in detail, instead the main arguments will be listed together with their strengths and weaknesses.

The first cluster of arguments centre on the manner in which statistics on population redistribution and change are interpreted. The most extreme position is that counterurbanisation is nothing real or new; it is instead an appearance produced by the failure of the administratively-defined 'urban' areas to encapsulate the continuously-expanding 'de facto' urban regions (Gordon, 1971; Burns, 1981; Hamnett, 1982; Pumain, 1983). Since much of the appearance of urban decline and of rural rejuvenation during the 1960s had been due to the statistical underbounding of urban areas during a period of rapid suburbanisation, it was inevitable perhaps that the trends in the 1970s would be viewed as a product of the same process now operating at an extended spatial scale.

There is, however, a serious problem with this argument. The mix of processes affecting the changes taking place within labour-market areas is very different from that affecting inter-labour-market area changes. The former are dominated by processes such as housing sub-market formation and social segregation; the latter are dominated by processes such as 'deindustrialisation' and the emergence of new spatial divisions of labour (see below). It is not wise to conflate the two levels or to gloss over their differences. Suburbanisation was the main form of population redistribution taking place within labour-market areas after 1950, but it slowed down appreciably during the early 1970s and in most western countries has remained at a low level ever since. At the inter-labour-market area level urbanisation was the main form of population redistribution until the late 1960s, and was replaced by counterurbanisation in the early-mid 1970s.

Although it is sensible and convenient to make this sharp distinction between the two levels, some qualifications are called for. The first is that decisions to locate investment to a new site within the labour-market area might be made on the basis of the same criteria as a decision to move to a new area. For example an insurance company seeking a more stable clerical workforce might choose to occupy office space in the suburbs of a metropolitan city rather than locate the investment in a provincial town. The second qualification arises from the fact that the definition of labour-market areas is not unproblematic. The labour-market areas of certain kinds of workers (such as female part-time employees) 'nest' within those of other kinds of workers, and the labour-market areas for those employed in the same type of work overlap in a complex fashion. A special problem arises with small and medium-sized towns located 50-100 kms from a metropolitan city; a small proportion (often a very small proportion) of the economically-active population (usually male white-collar workers) commute to the central city while most work locally or in nearby small towns. Many of these kinds of places have expanded in population and employment in the recent period but it would usually be unwise to assume that it is the commuting link with the major city that has been of central importance in that growth (Taaffe, 1980).

To summarise, although it is not always easy to draw the dividing line between local intra-labour-market area spatial deconcentration and non-local inter-labour-market area deconcentration, the statistical underbounding of urban areas is not the cause of the population 'turnround'. Statistical underbounding is indeed a problem in the analysis of migration data, and suburbanisation was important during the 1950-80 period, but the judgement that counterurbanisation replaced urbanisation around 1970 is based upon information at several spatial scales and however broadly defined the metropolitan and industrial areas are, the results are consistent with the view that more than just suburbanisation was involved. Suburbanisation might just be able to explain some of the variation in the net migration rates of Belgian provinces, but it is hardly equal to the task of explaining the variations in the rates of the Japanese prefectures (about twice as large in area), or the French regions (seven times larger), or the turnround in Irish net migration (twenty times larger), or that of the American South (one hundred times larger) (Fielding, 1982).

A second 'statistical' argument is that there is no inconsistency between a counterurbanisation pattern of net migration gains and losses and continued preference for large city living. This comes about through the fact that a very high proportion of the population of western countries live in the largest cities so that net migration gains to other areas at the expense of these cities can co-exist with a situation in which the inhabitants of these other areas are more likely to move to the cities than the inhabitants of the cities are to move to the other areas. In a country that had one city of one million inhabitants and 100,000 people living in rural areas the probability of a person moving from the city to the rural areas would have to be one tenth of the probability of a person moving in the reverse direction for their net migration balances to be zero (Alonso, 1978). While this is undeniably true its relevance to an explanation of counterurbanisation is restricted to those who place paramount importance upon migrant's preferences in their theoretical schemes (see below).

A related idea is that since the net migration gains of the major cities was fuelled by rural depopulation linked with the decline of agricultural employment, it was the drying up of the source of migrants that was responsible for the turnround (Myklebost, 1984). It is true that the proportion of the economically-active population employed in agriculture had fallen to a very low level in many western countries by 1970, but this proportion continued to decline after 1970 even if the numbers of people thus released to sustain urbanisation were much lower than those of the 1950s and 1960s. More importantly, the turnround in net migration was as marked a feature of regions with high rural population densities as of regions with low densities (Cencini et al., 1983).

Most of the remaining arguments of a 'statistical' nature address the problem of the reversal in major metropolitan area population change from growth to decline, and therefore emphasise factors such as falling birth rates since 1965 (Alonso, 1978) and changes in household composition and size (Bonvalet, 1983). These contributions and those of the urban life cycle school (Berg, 1981, 1982; Drewett, 1980, 1983; Hall and Hay, 1980; Klaassen, 1981) do not explain the trends in net migration and therefore will not be included in this review.

A second cluster of arguments centre on the notion that the changing patterns of net migration can best be understood by relating them to the place preferences of the migrants themselves. It is true that the counterurbanisation of the early 1970s coincided with a heightened sense of dissatisfaction with western urban life, powerfully expressed in the neo-Malthusianism of the 'limits to growth' school and in the disparate protest movements that came together under the banner of the 'green alliance' (Jones, 1984). It is also the case that circumstances can arise when individuals and families can exercise in a relatively unconstrained manner, a choice of a place of residence between different towns and regions. This is most obviously so for the wealthy retired (Cribier, 1982; Law,1976), but it can also happen when, in a highly segmented labour market, the supply of appropriately qualified persons for particular kinds of work is far less than the number of jobs on offer. This situation tends of course to arise most readily during times of general labour shortage such as existed in many western countries during the 1960s. However, even in times of high unemployment, the choice of place of residence can, for some, be relatively unconstrained. Some commentators on the very high rates of unemployment in certain resort areas, suggest that this is due to the attractiveness of these areas to the unattached unemployed (Dean et al., 1984).

Those who use these arguments found support in the results of questionnaire surveys of migrants' reasons for moving which showed that environmental considerations figured prominently (Long, 1979), and that the most preferred residential location was that of a small town within fairly easy travelling distance of a major city (Zuiches, 1980). They also point out that the opportunities for retirement migration had increased with the expansion of occupational pension schemes (Bourne, 1980), and that the attractiveness of such a move had been increased through familiarity with other regions as a result of heightened geographical mobility during career development, increased leisure travel and a rise in the ownership of 'second homes' (Coppock, 1977; White, 1984).

Nevertheless, despite the popularity of the place preference explanation among most North American and certain Western European writers (Long, 1979; Drewett, 1983), the contention here is that since migration decisions are typically constrained by wealth, employment, housing and family considerations, this position should play no more than a supportive role in the explanation of counterurbanisation and should be largely confined to tourist employment and retirement migration (Fielding, 1982).

A third cluster of arguments centres on economic changes that have advantaged the rural and peripheral regions over the metropolitan regions (Keeble, 1980). Some explanations within this category are tautological in that they are only trivially or definitionally true; their correctness lies in their mode of expression not in the world that they are supposed to represent. For example, to say that counterurbanisation has occurred because, after 1970, the diseconomies of agglomeration outweighed the economies, or that Myrdalian 'spread' effects came to predominate over 'backwash' effects is not to explain the development but to describe it in a different way. The question 'what happened to reverse the balance of advantage and disadvantage?' remains unanswered.

Similarly the standard (neo-classical) economic model of migration is not much help in explaining the turnround (Fielding, 1982; Gober, 1984). Of central significance to an assessment of the usefulness of economic theory is the fact that counterurbanisation involves net migration loss in those regions that retain high per capita incomes and low unemployment rates (the metropolitan regions), and net migration gain in the regions of low wages and high unemployment (the rural and peripheral regions) (Plane, 1984; Bouchet, 1983). This is precisely the opposite of what would be expected on the basis of the neo-classical economic model (Zelinski, 1983).

Another line of argument within this category is that the turnround has occurred because of economic recession and the energy crisis (Koch, 1980). US commentators point out that energy scarcity can have multiple effects on the urban and regional system (Wardwell and Brown, 1981). It can encourage investment in resource-rich regions located well away from the existing centres of industrial activity (such as northwest Australia, northern Scotland, west-central Canada and the west coast of Norway) and in this way assist the net migration turnround. On the other hand it should have the overall effect of discouraging deconcentration since it puts a premium upon the minimisation of transport costs and the achievement of spatial proximity (Long, 1981). The economic recession argument rests partly upon the experience of the 1930s when migration streams to the major industrial cities of the north-east and north-central USA were temporarily reduced by the very high unemployment rates that were to be found there. But the situation in the 1970s was interestingly different. In most western countries the development of social welfare services in the post World War II period meant that the unemployed were cushioned against the desperate poverty experienced by their pre-war equivalents. Furthermore, the decline in agricultural employment and even more so the concentration of agricultural land ownership in the hands of large capitalist farmers has resulted in a much reduced capacity of rural areas to absorb the urban unemployed. Thus, although much return migration to rural regions took place during the 1970s, very few of the migrants returned to agriculture (King, 1983; Rhoades, 1978). In any case, as many observers admit, the turnround came into being several years before the economic crisis of 1973/4; it occurred during the latter part of the long post-war period of growth and prosperity (Alonso, 1978; Inoue, 1983). It is perhaps pertinent to note that the recession argument is now coming to be used not as an explanation of the decentralisation tendencies of the 1970s, but of their subsequent demise! (Illeris, 1984).

Finally, there is the notion that the turnround in net migration was brought about by government policies for urban and regional development. Despite the fact that the new pattern of net migration correlated closely with the bias in these policies against the major metropolitan regions and in favour of the rural and peripheral regions, and despite the sizeable public funds devoted to regional development programmes notably in certain Western European countries, there is much evidence to suggest that the turnround would have occurred anyway; that is, without state policy assistance (Fielding, 1982). This judgement is supported by the facts (i) that the turnround occurred in those countries such as the US where area development programmes were poorly developed, and (ii) that where the policies were well-developed and were geared to the reversal of decline in both peripheral/rural and 'old industrial' regions such as France, it was only in the former that the turnround in net migration rates took place.

This brief review of most of the explanations presently available to us has proved to be rather negative in tone. Only the emphasis upon the role of place preferences in the migration decisions of individuals and households has been upheld with any real enthusiasm, and then with the proviso that it is only a minority of people in strictly limited situations that are in a position to act upon their preferences.

There is, however, one perspective on the causes of counterurbanisation that is, in the opinion of this reviewer, much more promising than those listed above. This perspective emphasises the importance of job-creating and job-destroying investment decision in determining population redistribution and trends in urbanisation, and it places these decision within a historical analysis of changing social class relations. The key concept is that of 'spatial division of labour' (Massey, 1979, 1984; Storper and Walker, 1984; Walker, 1985). It is asserted that the shift in the patterns of net migration towards a counterurbanisation form is best understood as resulting from the changeover from a situation in which each place concentrated upon all the tasks involved in the production of one or a related set of products or services (regional sectoral specialisation [RSS]), to one in which places became differentiated from one another with respect to the part or role they played in the production process (hierarchical/functional or simply 'new' spatial division of labour [NSDL]) (Hymer, 1972; Lipietz, 1982).

Up until the 1960s RSS had dominated, and because there was a common shift in 'western' societies away from employment in agriculture and towards the expansion of employment in (i) market-oriented mass-produced consumer goods and (ii) public and personal services, there was a major shift of population away from rural areas and 'old industrial' regions, and towards the largest industrial and commercial cities, especially when these were located in the 'core' regions of the national territory. This redistribution of population produced an 'urbanisation' relationship between net migration and settlement size (positive), and a net movement from the peripheral places to the core ones. Not surprisingly the normal method for analysing and forecasting urban and regional change during this period was 'shift-share' analysis. This applied national employment growth rates by sector to the employment structures of the individual places so as to produce the 'expected' employment change in those places. This was then compared with the actual employment change. A close fit was usually found between the expected change and the actual change, and this was interpreted to mean that the 'structural' component, that is, the effect due to favourable or unfavourable employment structure at the starting date was dominant (Ahnstrom, 1982).

By the 1970s it was clear that the structural component in shift-share analyses was no longer dominant; indeed in some studies it was found that the actual employment changes in places were inversely related to the changes expected on the basis of their sectoral employment structures (Fothergill and Gudgin, 1982). Thus rural areas which should, because of their agricultural employment, be declining were experiencing employment growth in both manufacturing and services, and the major metropolitan areas which should, because of their bias towards 'modern' industry and financial services, be growing, were in fact experiencing employment stagnation or slow decline overall, and rapid decline in their manufacturing employment. This puzzling situation is rather neatly explained by the notion that RSS had come to be replaced by the NSDL.

Under the NSDL, instead of the differences between places being due to the different parts they play in the social division of labour in society (a result of competition in the market), they arise from the separation of tasks within a particular production process (the technical division of labour which is a result of planning within the firm, or of a planned separation of tasks between firms). The most fundamental separation is that between conception/command and execution. Increasingly places were becoming differentiated on the basis of whether they were primarily oriented towards the activities of management, such as strategic decision-making about what goods or services should be produced, or primarily oriented towards the routine execution of tasks in factories, shops and offices (Scott, 1982). This differentiation did not take place in a vacuum; the emergence of a NSDL was closely related to the social histories of different areas; firms (and government agencies) evaluated places from the point of view of the firm's needs for different kinds of labour and from the point of view of their overall efficiency of operation. In general major metropolitan areas were judged to be necessary for the most specialised management functions and for those activities most closely linked to management such as financial services, banking and insurance, tax and other legal advice, and marketing services. Prestige environments near to or in easy contact with metropolitan cities tended to be chosen as sites for investment in research and development activities. Production requiring industrial skills, and for which a long tradition of craft labour was an advantage tended to gravitate towards the long-established industrial areas (often coalfield regions or major ports). Routine production was generally closed down in the major metropolitan and old industrial regions, and new investment took place in rural and peripheral regions (or abroad) or in places where there were workers that had not previously been employed (such as women in certain coalfield areas) (Clark, 1981). This separation of labour forces took place within organisations employing office staffs as well as those employing manual workers. The favoured locations for the routine tasks of the former were medium-sized provincial cities and free-standing towns close by the major metropolitan cities; the favoured locations for the latter were small and medium-sized towns in rural areas (Aydalot, 1984; Erickson, 1980; Todtling, 1984).

The reasons for the changeover from RSS to NSDL are many and complex, but they include the trend towards a concentration of ownership and control as multiplant and multiproduct companies became first national and then 'transnational' in the scope of their operations (Firn, 1975); the equalisation of the 'general conditions of production' between different places accompanying the standardisation of publicly-provided goods and services such as the road, rail, air and telecommunications networks, electricity, gas and water provision and health and education services (Wardwell, 1980); and the effects of technological change which led to a deskilling of many jobs as well as a reduction in the need for production of goods and services to be serviced by very large workforces concentrated in single workplaces. It might well be argued, however, that all of the above were 'facilitative' for the changeover to a NSDL in the sense that they made it possible for organisations to consider a spatial separation of functions and in particular a dispersal of routine production to rural and peripheral regions, but that there must have been reasons for such a dispersal. Indeed there were. The rapid and fairly continuous growths of the western economies after 1945 led to a general shortage of labour during the 1960s, and thus to recruitment and industrial-relations problems for employers with large workforces, working in large establishments in the major industrial and metropolitan cities. In situ substitution of machinery for labour was resisted and was often difficult to effect. One partial solution to this problem was the use of immigrant workers, and it was these workers that came to occupy many of the routine factory, building-site and other manual jobs in the major industrial cities from the early 1960s on. The other solution was to seek out 'green' labour elsewhere. This was to be found among women in rural and old industrial areas and more generally among young men and women in smaller labour market areas where the work practices and attitudes of the large factory or office were not endemic. These differences between towns and regions with respect to their social histories and political cultures (Bagnasco, 1981; Weinstein, 1978) were sometimes manifested in variations in levels of unionisation (many US studies emphasise the low levels of unionisation in the 'sunbelt' states in comparison with the industrial Northcentral and Northeastern states) (Sawers and Tabb, 1984; Peet, 1983), but they were also revealed in the incidence of industrial conflict and in levels of labour turnover, both of which tended to be higher in larger industrial cities than in smaller towns and rural areas.[3] The switching of investment towards these smaller places and more peripheral regions was accompanied by a major disinvestment, especially in manufacturing industry, in the largest cities (Ahnstrom, 1984; Bluestone and Harrison, 1982; CDP, 1977; Harrison, 1984; Miles, 1982; Townsend, 1983).

The changeover from RSS to NSDL might be expected to produce certain migration outcomes; these can be compared with what actually happened:

(i) it would be expected that overall levels of inter-labour market area mobility should decrease (a) because as sectoral diversification replaces sectoral specialisation the 'push factor' of job loss in the area's staple industry should be much reduced (Aydalot, 1983). For example, this diversification breaks the links between a decline in agricultural employment and rural depopulation; (b) because more generally, job creation in non-metropolitan labour markets would be expected to halt non-metropolitan to metropolitan migration; and (c) because as female activity rates increase so also do the problems of mobility for two-job households. As was seen in the introduction inter-labour market movements have indeed decreased during the period since about 1970.

(ii) it would be expected that the social composition of migration streams would change (a) because manual

[3]It is not being claimed that the switching of manufacturing and 'basic' service investment to small and medium sized towns and rural areas is only a response to these urban and regional differences in social history. Site development constraints, high land costs and burdensome taxation have contributed to the counterurban trend in location decisions taken by employers, and differences in housing costs favouring non-metropolitan areas have encouraged the same trend amongst owner-occupiers.

worker mobility would be less not only as a result of the reasons listed above but also because the increasing importance of the 'social wage', and the increasing standardisation of its various elements (health, education, social security...) between places reduces the advantages to be obtained from moving away from unemployment, poor job security or low wages; and (b) because to obtain an efficient use of their higher paid manpower, organisations would transfer technical and managerial personnel from one part of their operations to another. Again this is found to be the case. Transfers within organisations constituted about one quarter of all migration-decisions (of both the economically active and the retired) in the US in the 1970s (Long, 1979), and the 'new middle class' constitute a large and increasing proportion of all economically-active migrants (Davanzo, 1978; McKay and Whitelaw, 1978; Pohl and Soleilhavoup, 1982).

(iii) it would be expected that the manufacturing disinvestment and job loss in the major metropolitan and large industrial cities and the sizeable new investment in both manufacturing and services in small and medium-sized towns in rural and peripheral regions would lead at least to an end to the 'urbanisation' pattern of new migration gains and losses, and potentially to a reversal of that pattern. That this has indeed been so has been amply demonstrated in the first part of this paper. The NSDL concept, however, also suggests that, while de-industrialisation may well affect all large cities, those that house the highest-order functions of companies and public institutions may well sustain their employment levels through the further accretion of these kinds of activities (Goddard, 1978). There are some signs that this is the case, with capital cities and cities with major head office quarters maintaining their populations through migration, while other large industrial cities continue to lose theirs.

It is of course no proof of a set of ideas to say that what has happened is conformable with the outcomes predicted, but the relative lack of correspondence of expected and actual outcomes found in the other cases compared with the 'RSS to NSDL transition model' encourages further investigation of the significance of the changes in the 'geography of production' for an understanding of recent trends in migration and urban development in western countries.

In the late 1970s the counterurbanisation trend was itself modified (Archangeli, 1980). The earlier period of decentralised production based upon investments in branch-plants and relocated offices gave way somewhat to a new form of decentralised production based upon subcontractual relationships between 'core' firms, located in the metropolitan cities, and 'peripheral' small firms usually located in small and medium-sized towns in rural and peripheral regions (Aydalot, 1983b; Brusco, 1982; Danson, 1982; Murray, 1983; Norton, 1979). One example of this is the armaments industry. Government 'defence' establishments and associated manufacturing seem to have been important in explaining the rapid economic growth of parts of the American south (Tabb in Sawers and Tabb, 1984), and the high-technology equipment required for modern warfare formed the basis of the rapid expansion of 'Silicon Valley' in California (Saxenian in Markusen, 1983). Defence procurement and defence establishments are also involved in the economic growth of the 'M4 corridor' in Britain and of parts of the 'midi' in France. As western economies are further penetrated by the imports of consumer goods mass-produced in east and southeast Asia and in other newly-industrialised countries, it may well be that the emergence of these zones of small-scale production of high-technology products and related services will become the dominant element of counterurbanisation - a trend that began on the very different basis of 'Fordist' branch-plant mass-production of standardised products for known markets in the late 1960s and early 1970s (Massey, 1983; Sabel, 1982).

REFERENCES

Ahnstrom, L. (1982) 'The Concentration of a Compound - the Deconcentration of its Parts: the Economically-active Population of the Stockholm Region 1950-75', Geog.Ann. 64B, 69-75

Ahnstrom, L. (1984) The Turnround Trend and the Economically-active Population of Seven Capital Regions in Western Europe, Univ. Oslo, mimeo.

Alonso, W. (1978) 'The Current Halt in the Metropolitan Phenomenon', in C.L. Leven, (ed.), The Mature Metropolis, Heath, Lexington, Mass. pp. 23-41

Archangeli, F. et al. (1980) 'Patterns of Peripheral Development in Italian Regions 1964-77', Pap. and Proc. Reg. Sci. Assoc., 44, 19-34

Association of Japanese Geographers (1980) Geography of Japan, Teikoku-Shoin, Tokyo

Aydalot, P. (1983) 'France' in L.H. Klaassen and W.T.M. Molle (eds.), Industrial Mobility and Migration in the European Community, Gower, Aldershot, pp. 35-93

Aydalot, P. (1983b) The New Spatial Dynamisms in Western Europe: the French Case, mimeo, 31p

Aydalot, P. (1984) 'The Reversal of Spatial Trends in French Industry Since 1954' in J.G. Lambooy (ed.), New Spatial Dynamics and Economic Crisis. Internat. Reg. Sci. Ass., pp. 41-62

Bagnasco, A. (1981) 'Labour Market, Class Structure and Regional Formations in Italy', Int. J. Urb. Reg. Res., 5(1), 40-44
Beale, C.L. (1977) 'The Recent Shift of US Population to Non-Metropolitan Areas 1970-75', Int. Reg. Sci. Rev., 2, 113-122
Berg, L., van den and Meer, J., van der (1981) 'Urban Change in the Netherlands' in L.H. Klaassen (ed.), Dynamics of Urban Development, Gower, Aldershot, pp. 137-169
Berg, L., van den et al. (1982) Urban Europe: a Study of Growth and Decline, Pergamon, Oxford
Berry, B.J.L. (ed.) (1976) Urbanization and Counterurbanization, Sage, Beverly Hills
Berry, B.J.L. and Dahmann, D.C. (1980) 'Population Redistribution in the US in the 1970's' in B.J.L. Berry and L.P. Silverman (eds.), Population Redistribution and Public Policy, National Academy of Sciences, Washington D.C.
Bluestone, B. and Harrison, B. (1982) The Deindustrialization of America, Basic, New York
Bonvalet, C. and Lefebvre, M. (1983) 'Le Depeuplement de Paris 1968-75: Quelques Elements d'explication', Popu lation, 38(6), 941-958
Bouchet, J. (1983)'Mouvements Demographiques et Evolutions du Territoire', Rev. d'Econ. Reg. Urb., 5, 669-692
Boudoul, J. and Faur, J-P. (1982) 'Renaissance des Communes Rurales ou Nouvelle Forme d'urbanisation?', Econ. et Stat., 149, I-XVI
Bourne, L.S. (1980) 'Alternative Perspectives on Urban Decline and Population Deconcentration', Urb. Geog., 1(1), 39-52
Bourne, L.S. et al. (eds.) (1984) Urbanization and Settlement Systems, O.U.P., London
Brown, D.L. and Wardwell, J.M. (eds.), (1980) New Directions in Urban-Rural Migration: the Population Turnround in Rural America, Academic Press, London
Brusco, S. (1982) 'The Emilion Model: Productive Decentralization and Social Integration', Camb. J. Econ., 6(2), 167-184
Burnley, I.H. (1980) The Australian Urban System, Longman, London
Burns, L.S. (1981) 'The Metropolitan Population of the US: Historical and Emerging Trends' in Klaassen, L.H. (ed.), Dynamics of Urban Development, Gower, Aldershot, pp. 197-224
Carney, J. et al. (1983) L'Italia Emergente, Angeli, Milan
Champion, A.G. (1986) 'Population Trends in the 1970s', in J.B. Goddard and A.G. Champion (eds.), The Urban and Regional Transformation of Britain, Methuen, London

Clark, G.L. (1981) 'The Employment Relation and the Spatial Division of Labour: A Hypothesis', Ann. Ass. Amer. Geogs., 17(3), 412-424
CDP (Community Development Project) (1977) The Costs of Industrial Change, CDP, London
Coppock, J.T. (ed.) (1977) Second Homes: Curse or Blessing?, Pergamon, Oxford
Courgeau, D. (1978 and 1982) 'Les Migrations Internes en France de 1954 a 1975', Population, 33(3), 525-545 and 37(2), 341-370
Court, Y. (1984) Counterurbanization: A Review and Bibliography, Portsmouth Poly. mimeo, 40p
Cribier, F. (1982) 'Aspects of Retired Migration from Paris' in A.M. Warnes (ed.), Biographical Perspectives on the Elderly, Wiley, Chichester, pp. 111-138
Danson, M.W. (1982) 'The Industrial Structure and Labour Market Segmentation: Urban and Regional Implications', Reg. St., 16(4), 255-265
Davanzo, J. (1978) US Internal Migration: Who Moves and Why? US House of Representatives, House Select Committee on Population, mimeo, 14p
Dean, K.G. et al. (1984) 'Counterurbanization and the Characteristics of Persons Migrating to West Cornwall', Geoforum, 15(2), 177-190
Drewett, R. (1980) 'Changing Urban Structures in Europe', Ann. Amer. Ac. Pol. Soc. Sci., 451, 52-75
Drewett, R. (1983) Towards Reurbanization and Deconcentration: the Duality in European Urban Development Trends in the Decade to 1981, LSE, mimeo, 29p
Erickson, R.A. (1980) 'Corporate Organization and Manufacturing Branch-Plant Closures in Non-Metropolitan Areas', Reg. St., 14(6), 491-502
ERIPLAN (1978) 'Urban Development in Sweden 1960-75: Population Dispersal in Progress' in N.M. Hansen (ed.) Human Settlement Systems, Ballinger, Cambridge, Mass., pp. 51-83
Fielding, A.J. (1966) 'Internal Migration and Regional Economic Growth: A Case Study of France', Urb. St., 3(3), 200-214
Fielding, A.J. (1982) 'Counterurbanization in Western Europe', Prog. in Plan., 17(1) 1-52
Fielding, A.J. (1984) Trends in Urban Development in Western Europe 1950-80, Transport and Road Research Laboratory WP/SRB 26, mimeo, 91p
Firn, J., (1975) 'External Control and Regional Development: The Case of Scotland', Env. Plan., A7(4) 393-414
Fothergill, S. and Gudgin, G. (1982) Unequal Growth: Urban and Regional Employment Change in the UK, Heinemann, London
Francart, G. (1983) 'Le Reequilibrage Demographique de la France', Econ. et Stat., 153 35-46

Gober, P. (1984) 'Regional Convergence Versus Uneven Development: Implications for Sunbelt SMSA's', Urb. Geog., 5

Goddard, J.B. (1978) 'Changes in Corporate Control in the British Urban System 1972-1977', Env. Plan., A10(9), 1073-1084

Goldstein, S. and Sly, D.F. (eds.) (1975) Patterns of Urbanization, Ordina, Dolhain, 2 vols

Gordon, P. (1979) "Deconcentration without a 'Clean Break'", Env. Plan., A11(3), 281-290

Greenwood, M.J. (1981) Migration and Economic Growth in the US: National, Regional, and Metropolitan Perspectives, Academic Press , New York

Hall, P. and Hay, D. (1980) Growth Centres in the European Urban System, Heinemann, London

Hall, P. (1983) 'Decentralization Without End? A Re-evaluation' in J. Patten (ed.) The Expanding City, Academic Press, London, pp. 125-155

Hamnettt, C. and Randolph, W. (1982) 'The Changing Population Distribution of England and Wales 1961-81: A Clean Break or Consistent Progression?', Built Env., 8(4), 272-280

Harrison, B. (1984) "Regional Restructuring and 'Good Business Climates'" in L. Sawers and W. Tabb (eds.), Sunbelt/Snowbelt, Oxford UP, London, pp. 48-96

Hauser, P.M. (1981) 'The Census of 1980', Sci. Amer., 245(5), 53-61

Hugo, G. (1983) Interstate Migration in Australia 1976-81, Flinders University, mimeo, 68p

Hugo, G. (1984) Internal Migration in Australia: Some First Glimpses from the 1981 Census, Flinders University, mimeo

Hymer, S. (1972) 'The Multinational Corporation and the Law of Uneven Development' in J.N. Bhagwati (ed.), Economics and World Order, Free Press, New York, pp. 113-140

Illeris, S. (1979) 'Recent Development of the Settlement System of Advanced Market Economy Countries', Geog. Tidss., 78(9), 49-56

Illeris, S. (1981) Research On Changes in the Structure of the Urban Network, AKF, Copenhagen

Illeris, S. (1984) 'Danish Regional Development During Economic Crisis', Geog. Tidss., 84(1), 53-62

Inoue, S. (1983) 'Stagnant Growth of Japanese Major Metropolitan Regions in the Era of Post-Industrial Development' in P.A. Morrison (ed.), Population Movements: Their Forms and Functions in Urbanization and Development, Ordina, Liege, pp. 211-237

Jarvie, W.K. (1984) The Turnround in Australia, D. Phil Thesis, Flinders University

Jones, H.R. et al. (1984) Peripheral Counterurbanization (N. Scotland), Dundee University, mimeo, 28p

Kawabe, H. (1980)'Internal Migration and the Population Distribution in Japan' in Japanese Association of Geographers, Geography of Japan, Teikoku-Shoin, Tokyo, pp. 379-389

Kawashima, T. (1980) 'The Regional Pattern of the Japanese Economy' in Japanese Association of Geographers, Geography of Japan, Teikoku-Shoin, Tokyo, pp. 390-414

Keeble, D. (1980) 'Industrial Decline, Regional Policy and the Urban-Rural Manufacturing Shift in the UK', Env. Plan., A12, 945-962

Keeble, D. (1983) 'The Urban-Rural Manufacturing Shift in the European Community', Urb. St., 20, 405-418

King, R. and Strachan, A. (1985) 'Back to Bernalda: The Dynamics of Return Migration to a Southern Italian Agro-Town' in G.A. van der Knaap and P.E. White (eds.), Contemporary Studies in Migration, Geobooks, Norwich

Klaassen, L.H. and Scimemi, G. (1981) 'Theoretical Issues in Urban Dynamics' in L.H. Klaassen (ed.) (1981) Dynamics of Urban Development, Gower, Aldershot, pp. 8-28

Koch, R. (1980) 'Counterurbanization' auch in Westeuropa', Inf. zur Raument., 2(2), 59-69

Kontuly, T. et al. (1984) Counterurbanization in the Federal Republic of Germany, University of Utah, mimeo, 26p

Kuroda, T. (1975) 'Urbanization and Population Redistribution in Japan' in S. Goldstein and D. Sly (eds.), Patterns of Urbanization, Ordina, Dolhain, pp. 433-463

Law, C.M. and Warnes, A.M. (1976) 'The Changing Geography of the Elderly in England and Wales', Trans. Inst. Brit. Geogs., NS1(4), 453-471

Lipietz, A. (1980) 'The Structuration of Space, The Problem of Land and Spatial Policy' in J. Carney et al., Regions in Crisis, Croom Helm, London, pp. 60-75

Long, J.F. (1981) Population Deconcentration in the US, US Bureau of the Census, CDS-81-5, 105p

Long, L.H. and Hansen, K.A. (1979) Reasons for Interstate Migration, US Bureau of the Census, P-23(81), 31p

Long, L.H. and De Are, D. (1982) 'Repopulating the Countryside: A 1980 Census Trend', Science, 217(4565), 111-116

McKay, J. and Whitelaw, J.S. (1978) 'Internal Migration and the Australian Urban System', Prog. in Plan., 10(1), 1-83

Markusen, A.R. (ed.) (1983) 'Silicon Landscapes: High Technology and Job Creation', Special issue of Built Env., 9(1), 1-78

Massey, D. (1979) 'In What Sense a Regional Problem?', Reg. St., 13(2), 233-243

Massey, D. (1983) 'The Shape of Things to Come', Marxism Today, April 1983, pp. 18-27

Massey, D. (1984) Spatial Divisions of Labour, Macmillan, London
Miles, N.J.O. (1982) The Spatial Decentralisation of Manufacturing Growth and the Restructuring of the Labour Process, LSE, mimeo, 45p
Morrison, P.A. (ed.) (1983) Population Movements: Their Forms and Functions in Urbanization and Development, Ordina, Liege
Murray, F. (1983) 'The Decentralization of Production - The Decline of the Mass-Collective Worker?', Capital and Class, 19, 74-99
Myklebost, H. (1984) The Evidence for Urban Turnround in Norway, Geoforum, 15(2), 167-176
Nanjo, Z. et al. (1982) Migration and Settlement: 13 Japan, IIASA, RR-82-5
Norton, R.D. and Rees, J. (1979) 'The Product Life Cycle and the Decentralisation of American Manufacturing', Reg. St., 13(2), 141-151
Ogilvie, A.A. (1979) 'Migration - The Influence of Economic Change', Futures, October 1979, pp. 383-394
Ogilvie, A.A. (1982) 'Population Migration Between the Regions of GB', Reg. St., 16(1), 65-73
Peet, R. (1983) 'Relations of Production and the Relocation of US Manufacturing Industry Since 1960', Econ.Geog., 59(2), 112-143
Plane, D.A. and Isserman, A.M. (1983) 'Interstate Labor Force Migration: An Analysis of Trends, Net Exchanges and Migration Subsystems', Socio-Econ. Plan. Sci., 17(5-6), 251-266
Plane, D.A. (1984) US Core-Periphery Net Migration, University of Arizona, mimeo, 25p
Pohl, R. and Soleilhavoup, J. (1982) Mobilite Professionnelle, INSEE Coll. D91 Paris
Pumain, D. (1983) Deconcentration Urbaine, Population et Societes, 116, Feb. 1983
Rhoades, R.E. (1978) 'Intra-Urban Migration and Rural Development: Lessons from the Spanish Case', Hum. Org., 37, 131-147
Robert, P. and Randolph, W.G. (1983) 'Beyond Decentralisation: The Evolution of Population Distribution in England and Wales 1961-81', Geoforum, 14(1), 75-102
Roseman, C.C. (1977) 'Changing Patterns Within the US', Ass. Amer. Geogs., Washington DC, 37p
Sabel, C.F. (1982) Work and Politics: The Division of Labour in Industry, Cambridge University Press, London
Sawers, L. and Tabb, W. (eds.) (1984) Sunbelt/Snowbelt: the Political Economy of Urban Development and Regional Restructuring, Oxford University Press, London
Scott, A.J. (1982) 'Production System Dynamics and Metropolitan Development', Ann. Ass. Amer. Geogs., 72(2), 185-200

Storper, M. and Walker, R. (1984) 'The Spatial Division of Labor: Labor and the Location of Industries' in L. Sawers and W. Tabb (eds.), Sunbelt/Snowbelt, Oxford University Press, London, pp. 19-47
Taaffe, H.L. et al. (1980) 'Extending Commuting and the Intermetropolitan Periphery', Ann. Ass. Amer. Geogs., 70(3), 313-329
Todtling, F. (1984) 'Organizational Characteristics of Plants in Core and Peripheral Regions in Austria', Reg. St., 18(5), 397-412
Townsend, A.R. (1983) The Impact of Recession, Croom Helm, London
Vernon, R. (1963) Metropolis 1985, Anchor, New York
Vining, D.R. and Strauss, A. (1977) 'A Demonstration that the Current Deconcentration of Population in the US is a Clean Break with the Past', Env. & Plan., A9, 751-58
Vining, D.R. and Kontuly, T. (1978) 'Population Dispersal from Major Metropolitan Regions: an International Comparison', Int. Reg. Sci. Rev., 3(1), 49-73
Vining, D.R. and Pallone, R. (1982) 'Population Dispersal from Core Regions: a Description and Tentative Explanation of the Patterns in 22 Countries', Geoforum, 13(4) 339-410
Walker, R.A. (1985) 'Class, Division of Labour and Employment in Space' in D. Gregory and J. Urry (eds.), Social Structure and Spatial Relations, Cambridge University Press, London
Wardwell, J.M. and Brown, D.L. (1980) 'Population Redistribution in the US During the 1970's' in D.L. Brown and J.M. Wardwell (eds.), New Directions in Urban-Rural Migration, Academic, New York, pp. 5-35
Wardwell, J.M. (1980) 'Toward a Theory of Urban-Rural Migration in the Developed World' in D.L. Brown and J.M. Wardwell (eds.), New Directions in Urban-Rural Migration, Academic Press, New York, pp. 71-114
Weinstein, B.L. and Firestone, R.E. (1978) Regional Growth and Decline in the US: the Rise of the Sunbelt and the Decline of the Northeast, Praeger, New York
White, P. (1984) The West European City: a Social Geography, Longman, Harlow
Whitelaw, J.S. et al. (1984) 'The Australian Urban System' in L.S. Bourne, et al. (eds.), Urbanization and Settlement Systems, Oxford University Press, pp. 71-91
Witherick, M.E. (1983) 'Recent Population Changes in Japan and their Spatial Implications', Geog., 68(2), 97-114
Yamaguchi, T. (1984) 'The Japanese National Settlement System' in L.S. Bourne, et al. (eds.), Urbanization and Settlement Systems, Oxford University Press, London, pp. 261-279
Zelinski, W. (1971) 'The Hypothesis of the Mobility Transition', Geog. Rev., 61, 219-249

Zelinski, W. (1983) 'The Impasse in Migration Theory: a Sketch-Map for Potential Escapees' in P.A. Morrison (ed.), Population Movements: Their Forms and Functions in Urbanization and Development, Ordina, Liege, pp. 19-46

Zuiches, J.J. (1980) 'Residential Preferences in Migration Theory' in D.L. Brown and J.M. Wardwell (eds.), New Directions in Urban-Rural Migration, Academic Press, New York, pp. 163-188

Chapter Nine

MIGRATION AND INTRA-URBAN MOBILITY

M. Cadwallader

The study of migration represents a truly inter-disciplinary field of endeavour, involving geographers, economists, sociologists, demographers, and others. Although each discipline projects a particular orientation, such that economists have tended to emphasise the interrelationships between labour markets and migration while geographers have highlighted the spatial structure of inter-regional flows, there has been a genuine convergence of the disciplinary foci. The present essay reflects this convergence in that the concepts and models have been culled from a variety of perspectives. As in most recent reviews of the migration literature (Clark, 1982), a distinction is made between inter-regional migration and migration within cities, or residential mobility. Within these two broad categories the reader will also discern a somewhat more informal distinction between macro and micro-level approaches.

INTER-REGIONAL MIGRATION

Many of our current ideas about inter-regional migration stem from the pioneering work of Ravenstein (Grigg, 1977). Ravenstein postulated a set of "laws" or, more accurately, generalisations that were inductively derived from place of birth data for various parts of the world. In particular, he made statements concerning both the characteristics and determinants of population movement. For example, two of his generalisations suggest that most migrants move only a short distance, and that the major causes of migration are economic. Many of his original hypotheses have since been confirmed, and the relative importance of push and pull factors continues to provide a theoretical framework for much contemporary research (Dorigo and Tobler, 1983).

Spatial Patterns

More recent work on the spatial pattern of migration flows has emphasised both distance and directional biases. In terms of distance, a variety of studies have shown that the number of migrants from any given region decreases with increasing distance. In a now classic study, Hagerstrand (1957) examined the relationship between migration and distance for the village of Asby in Sweden. He concluded that migration is indeed an inverse function of distance, although the temporal decline in the distance coefficient indicated that Asby's migration field was expanding. Since Hagerstrand's study a number of scholars have investigated the functional form of the relationship linking migration and distance, and have suggested a range of possible alternatives, including the double-log, negative exponential, and square-root exponential (Haynes, 1974).

Regardless of the precise functional form, however, the weight of evidence indicates that the role of distance has decreased over time. For example, Clayton (1977) found that the contribution of distance to the total explained variation in inter-state migration in the United States declined progressively between 1935 and 1970. There is considerable doubt, however, as to whether the individual effect of distance can ever be satisfactorily isolated. It has been argued that distance-decay parameters reflect a complex combination of spatial structure, involving the size and configuration of origins and destinations in a spatial system, and intrinsic interaction behaviour, thus precluding any simple interpretation of these parameters (Fotheringham, 1981).

With respect to directional bias, the results of descriptive studies are perhaps even more difficult to interpret. When examining the directional bias in migration flows between cities in the United States, Wolpert (1967) concluded that there was a distinct bias toward the South-West. As with distance, however, such a conclusion is tempered by the fact that the directional components of migration streams are inevitably constrained by the existing geometry of population distribution. The exploration of migration flows through various forms of linkage analysis seems potentially more promising, and this approach has been applied to migration data from various parts of the world. On the basis of interaction patterns, the linkage analysis generates a hierarchical system of migration fields (Slater, 1984), the changing nature of which can be usefully monitored. For example, using United States data for 1935 to 1970, Clayton (1977) revealed a reduction in the number of terminal nodes, reflecting an increasing concentration around a few major destinations, such as California, Florida, and Texas.

In addition to identifying distance and directional biases, a variety of authors have attempted to study migration via the popular gravity model. This model postulates that the amount of migration between any two regions, or cities, will be directly proportional to the product of their populations, and inversely proportional to the distance between them. If the variables are expressed in logarithmic form, multiple regression analysis can be used to estimate the coefficients associated with population and distance. Although criticised for its rather weak theoretical base, the gravity model served to demonstrate how the notion of push and pull factors might be profitably operationalised. In particular, rather than simply using population as a measure of regional attractiveness, a variety of other variables began to be substituted into the equation. For example, along with distance, Lowry (1966) used number of persons in non-agricultural employment, manufacturing wage rates, and unemployment. When fitting a modified form of Lowry's model to inter-metropolitan migration flows in California, Rogers (1967) found that over 90% of the variation was accounted for. Since then a long list of authors have sought to identify the relative importance of various economic and social factors through the construction of single-equation regression models.

Single-Equation Models

An extensive list of variables have been used in such single-equation migration models (Greenwood, 1975). First, most research has shown that unemployment rates tend to be positively related to out-migration and negatively related to in-migration. Second, theories rooted in equilibrium economics argue that in any situation characterised by interregional wage differentials, labour will tend to migrate from low to high income regions. A number of studies have demonstrated that in-migration is positively related to income levels, although it appears that destination income has a greater influence on migration flows than origin income. Third, educational differences account for different migration rates between places, with the well-educated being more likely to be involved in long distance moves than their more poorly educated counterparts. Fourth, the propensity of labour force members to migrate is likely to decrease with increasing age, as older people have a shorter expected working life over which to realise the advantages of migration. Fifth, growth in non-agricultural employment tends to be positively related to levels of in-migration, although recently the growth rates for many non-metropolitan areas have exceeded those of metropolitan areas. Sixth, various studies have attempted to measure the impact of such environmental factors as climate.

Information about previous migration flows is also often included in migration models. For example, migrant stock, or the proportion of the population of the sender area residing in the receiver area, has been found to have a positive relationship with migration. This variable is generally included as a means of calibrating information levels, and implies that previous out-migrants send back information concerning employment opportunities and wage rates. Kau and Sirmans (1977) corroborate this line of reasoning when they argue that the migrant stock factor is especially important when explaining the behaviour of those migrants making their first inter-state move. Amount of migration in the previous decade has also been used as a measure of migrant stock, and found to be positively related to migration flows. This particular method of measurement, however, highlights one of the problems involved in interpreting the theoretical significance of the migrant stock variable, as one would expect migrant stock to be highly correlated with the current determinants of migration. As Greenwood (1970) suggests, one could perhaps disentangle the implications of migrant stock by comparing the results of different models in which it is included and excluded.

As one might imagine, the explanatory variables in such models are often interrelated, thus posing problems of multicollinearity when estimating the regression coefficients by ordinary least-squares analysis. Consequently, it has been suggested that the set of explanatory variables might first be subjected to principal components analysis, thus generating a smaller number of orthogonal components that reflect the underlying similarities of the original variables (Riddell, 1970). These components, as measured by the component scores, are then used in the regression equation, permitting a more realistic interpretation of the coefficients. Rodgers (1970) used this strategy when looking at the relationship between migration and industrial development in southern Italy. A principal component representing "socio-economic health" proved to be a good predictor of out-migration between 1952 and 1968.

Besides the technical problem of multicollinearity, single equation migration models also suffer from even more important theoretical defects. In particular, they ignore the possibility of two-way interaction between migration and the explanatory variables, even though it can be taken as axiomatic that although regional characteristics undoubtedly influence migration patterns, the reverse is also equally true. In other words, migration is a cause of economic change as well as a reaction to it. Regional development theory suggests that migration tends to increase income differences across regions, and empirical evidence supports this notion. Migration also influences unemployment levels, and out-migration can often worsen rather than improve the unemployment problems in depressed areas. Unemployed persons take with them a certain level of expenditure when they migrate, thus generating a set of multiplier effects that further decrease the level of employment in the region.

This interdependency between the characteristics of places and the flows between them has been investigated within the framework of General Field Theory. Field theory was originally developed as a method for analysing causal relations in social psychology. The methodology was more formally articulated by Rummel (1965), who divided the characteristics of social units, such as groups or nations, into attributes and types of behaviour. Berry (1966), in his work on Indian commodity flows, provided a more explicitly geographical perspective by examining the characteristics and flows between a set of regions. Within the context of migration, Schwind (1975) adopted a field theory framework to investigate inter-regional migration in the United States. More specifically, canonical correlation analysis was employed to unravel the interdependencies between migration behaviour and regional characteristics.

Canonical correlation analysis involves articulating the interrelationships between two groups of variables, in this case the regional characteristics on the one hand and the migration patterns on the other. The two sets of variables are transformed into orthogonal canonical vectors, such that the correlations between certain variables of the two sets are maximised. The canonical weights express the degree of association between the original variables and the canonical vectors, while the canonical correlation expresses the degree of association between the two sets of variables, for each canonical vector. In general, however, canonical correlation would only appear to be a useful tool where the investigator has little theoretical knowledge about the system at hand, and even in those situations where it is used as a primarily exploratory technique certain problems are encountered. First, it is often extremely difficult to substantively interpret the canonical vectors, as the importance of these vectors is not determined by how well they account for the variation within the two sets of variables, but rather how well they account for the relationships between the two sets. Second, the canonical weights cannot be interpreted like regression coefficients, as they show the degree to which the variables and canonical vectors are related, but not how they are related.

Structural Equation Models

A more suitable approach for investigating these complex interrelationships is available in the form of structural equation models, where the term is used to include both causal models, or path analysis, and systems of simultaneous equations. Such models not only specify the relationships between the independent variables and the ultimate dependent variable, in this case migration, but also explicate the interrelationships among the causally prior variables themselves. In particular, causal models and path analysis encourage the identification of direct and indirect effects between variables. For example, an increase in the minimum wage rate might directly increase the amount of migration to a region, but at the same time it might also increase the level of unemployment in that region, thus indirectly decreasing the amount of in-migration. When looking at net migration rates in the Upper Mid-West of the United States, Cadwallader (1985) was able to demonstrate that the direct effects of many variables on migration are at least partially offset by the indirect effects.

When significant feedback effects or reciprocal causation are present in the causal system it is appropriate to specify the interrelationships in the form of a simultaneous-equation model. Greenwood (1981) used nine equations and five identities when exploring rates of in-migration and out-migration for large cities in the United States. Besides measures of migration, the constituent variables included income change, manufacturing employment change, government employment change, and unemployment change. The various endogenous variables were found to be significantly interrelated, with in-migration inducing greater employment growth. In a similar study of inter-state migration, Gober-Meyers (1978) used a set of three simultaneous equations. Within this causal system net migration, migration benefits, and per capita income growth were expressed as both dependent and independent variables.

Most of these simultaneous-equation models have stressed the reciprocal relationships between migration, income, and employment. Using labour market areas in Sweden, Dahlberg and Holmlund (1978) were able to show that migration from a region tends to decrease income and employment growth in the sending region, although the opposite was true of in-migration. Similarly, Cadwallader (1985) constructed a simultaneous-equation model in which income, unemployment, and migration acted as the three endogenous variables. The structural equations were estimated using two-stage least-squares analysis, with agricultural employment, education, and urbanisation as exogenous variables. Income and migration were reciprocally related; high-income levels attracted migrants, but as net migration increased income levels decreased.

Dynamic Models

Structural equation models have often incorporated change over time by the use of lagged variables. Temporal change has received more formal attention, however, in the context of Markov Chain models. Migration transition probability matrices can be constructed to describe the proportion of people who, during a given time period, move from one region to another. Although Markov models are attractive in that the transition probabilities reflect both the relationships between regions and the stochastic nature of migration decisions, they are limited by a number of rather restrictive assumptions. First, the transition probabilities are assumed to be temporally stable. Second, the population is assumed to be homogeneous, implying that everyone obeys the same transition matrix. Third, the Markov property suggests that the probability of migrating between two areas is solely dependent upon current location, and not upon previous behaviour. Recent approaches have attempted to construct models of change that relax the assumption of stationary transition matrices by explicitly incorporating heterogeneous populations and changing preferences (Huff and Clark, 1978a). The lack of behavioural content in Markov models indicates that they might be profitably combined with the structural equation approach, which explicitly incorporates a set of explanatory variables (Rogerson, 1984).

An alternative approach for capturing the dynamics of migration flows is also available in the form of various time-series techniques. More specifically, one can develop moving average models and autoregressive models. In the moving average model the value of an individual variable, at a particular point in time, is generated by the weighted average of a finite number of previous and current random disturbances, or shocks. By contrast, in an autoregressive model the current observation is generated by a weighted average of past observations, together with a random disturbance in the current period. Mixed autoregressive-moving average models can be developed for stationary time series, and certain types of nonstationary time series can be differenced so as to produce a stationary time series, thus allowing the development of a general integrated autoregressive-moving average model. Within the context of migration this autoregressive integrated moving average process, known as ARIMA, suggests that the amount of migration from one time period to the next is a complex function of previous migration plus current and previous random disturbances.

The implications of these time-series models for migration have only just begun to be explored. Markovian models of interregional migration are essentially autoregressive. Similarly, demographic forecasting models that specify the determinants of population change by accounting for births, deaths, and migration (Rees and Wilson, 1977) are also autoregressive and assume the random disturbance term to be negligible. Most geographic processes are stochastic rather than completely deterministic, however, as a variety of unknown variables, or an exogenously generated random element, may affect the dependent variable. Consequently, when a variable like migration is treated in dynamic terms we are looking at an underlying stochastic process that can be treated as some form of ARIMA model.

Clark (1982a) has investigated gross migration flows using ARIMA models and suggests that both autoregressive and moving average processes are at work. In particular, Markovian, or autoregressive models, are apparently inappropriate for many rapidly growing regions. Such regions contain a sizeable moving average component, implying that in-migration is a volatile process, more influenced by exogenous and rapid shocks than by previous patterns of in-migration. In contrast, average and decelerating growth areas seem better approximated by autoregressive models. Future efforts along these lines are likely to incorporate spatial autocorrelation by using space-time autocorrelation functions. However, the prior assignment of weights reflecting the effects of contiguous interregional migration flows remains problematical. The usual assignment of weights according to distance-decay principles might be unwise, as regions are not necessarily highly integrated with their immediate neighbours, at least in the context of migration (Clark, 1982b).

Behavioural Approaches

A preliminary step towards formulating a more behaviourally oriented approach to migration involves the investment in human capital theory of migration, whereby a move from i to j is assumed to depend on the income differential between regions i and j, discounted for future income in i, less the costs of migrating from i to j. Migration is regarded as an investment because the benefits can only accrue over a period of time, and as the investment is in the individual or family it represents an investment in human capital. A cost-benefit framework is utilised, in which both financial and non-financial factors can be included. For example, the costs of migrating from i to j might involve "opportunity costs", expressing the loss of wages due to moving and searching for work, and "psychic costs", representing the psychological trauma of uprooting the household. More recent versions of the human capital theory have emphasised the notion of expected income differentials (Harris and Todaro, 1970), thus arguing that migrants attempt to maximise expected utility. Speare (1971) warns, however, that although the cost-benefit framework provides a reasonable representation of the factors involved in the migration process, it should not be implied that the expected costs and benefits are actually calculated. Many migrants will only have a rather vague idea concerning expected earning potential, and moving costs can often only be approximately estimated.

Despite its widespread use in the context of labour migration, the human capital approach has not been immune from criticism (Blaug, 1976). In particular, most empirical calibrations of the theory have used simple logit or probit transformations in which the decision to move is viewed dichotomously. Also, recent empirical evidence suggests that only comparatively small monetary rewards are obtained from the investment in migration (Grant and Vanderkamp, 1980). Attempts to expand the theory have included not only expected income differentials, but also amenity differentials and the anticipated benefits from local government services (Cebula, 1980). It has often been suggested that migrants will move to regions that best satisfy their preferences for public goods, and Kleiner and McWilliams (1977) argue that non-whites, in particular, are attracted to states with high levels of welfare benefits. A search theoretic framework can also be used to strengthen human capital theory, with due consideration being given to the flows of job information and aspects of job competition (Pickes and Rogerson, 1984).

Although the investment in human capital models of migration are formulated in terms of individual utility maximisation, they are frequently estimated using aggregate data that refer to average income and unemployment levels in the origin and destination regions. A more explicitly micro-level approach involves the concept of place utility, which summarises an intended migrant's attitude towards a potential destination (Wolpert, 1965). Thus, place utility represents the overall attractiveness of a particular place, for a particular individual, as evaluated across a set of characteristics. An individual's place utility matrix contains attributes as the rows and places as the columns, with the overall utility associated with any place being a weighted sum of the values in a column. Individual variations will occur due to the differential weighting of attributes and differential levels of information about places. More specifically, the ranking of attributes will reflect stage in the life cycle, while information levels appear to be a function of size and distance variables (Cadwallader, 1978).

Lieber (1978) has attempted to uncover the attributes of places which influence their associated utility values by using the grid-sorting technique of personal construct theory and the semantic differential technique. The grid-sorting technique allows subjects to directly identify the relevant attributes of stimuli, in this case places, that are being compared. By contrast, the semantic differential technique involves presenting the subjects with a set of stimuli which they are required to evaluate across a series of scales consisting of bipolar adjectives. Some form of factor analysis is then used to uncover the underlying dimensions, or cognitive categories. Using these methods, Lieber found that four major variables best characterise the evaluative features of potential destinations: proximity to a major city, proximity to fresh air recreational opportunities, proximity to close relatives, and general economic conditions.

Metal Maps and Migration

A growing number of researchers have begun to explore ways of using the subjective evaluation of alternatives as a theoretical framework for constructing migration models. Much of the work in this area was originally inspired by Gould's (Gould and White, 1974) pioneering investigations of mental maps, in which he explored the perceived residential desirability of different regions within the United States and Britain. Isolines were used to connect points of equal value, thus creating a preference surface that reflected the hills and valleys of desirability for a particular group of people. Since then efforts have been made to measure the degree of association between these mental maps and migration flows.

Lloyd (1976) argued that although it has been suggested for some time that spatial behaviour is at least partly a function of an individual's cognitions of his or her alternatives, no attempt has been made to explicitly examine the linkages among cognition, preference, and behaviour. With this thought in mind, he identified three abstract spaces that could be used to analyse inter-state migration flows. First, an n-dimensional cognitive space contained information concerning the cognised characteristics of places. The axes represent the underlying cognitive structure of the attributes associated with those places, and so the cognised characteristics of any particular place are defined by its location within that space. Second, an n-dimensional preference space contained axes which represent the underlying preference structure, and define the degree to which a particular place is preferred. Third, an n-dimensional behaviour space, such as that associated with inter-state migration, involved axes that represent the underlying structure of actual behaviour.

Within this conceptual framework, data were gathered from samples of university students. The grid repertory test was used to collect the cognitive information, and multidimensional scaling was used to identify the underlying dimensions of the cognitive spaces. Similarly, Gould's rank-ordering technique was utilised to collect the preference data, with multidimensional scaling again determining the structure of the preference spaces. Finally, the behaviour space was generated using a 48 by 48 origin-destination matrix, and subjecting it to the revealed preference procedures used by Rushton (1981). A series of canonical correlation analyses suggested strong linkages between cognitions, preferences, and actual behaviour. Lloyd ended by noting, however, that despite these encouraging results, considerably more work was needed in terms of formally explicating the relationships between the cognitive constructs and actual behaviour, and also in terms of identifying how the cognitive information is translated into overall preferences for particular destinations, which are then translated into overt behaviour.

Using twenty-five Kentucky cities, White (1974) also focused on the relationship between preferences and overt behaviour. He provided evidence to support the hypothesis that the in-migration to a particular city is related to the preference value attributed to that city by a sample of potential migrants. Indeed, the correlation between residential preference and in-migration indicated that almost half the variation in migration behaviour can be accounted for by preferences. In a later paper, however, using the same data set, White (1978) cautioned that models linking aggregated preference maps with aggregated behaviour are limited by the fact that preference patterns vary across different socio-economic groups. This problem is compounded by the fact that socio-economic groups vary in terms of their propensity to migrate. White concluded by suggesting that the preference of potential migrants might be usefully disaggregated on the basis of accessibility to alternative destinations, socio-economic status, and awareness levels for alternative locations.

In another study, using a data set involving metropolitan areas throughout the United States, White (1980) showed that out-migration from Topeka, Kansas to a selection of other cities was more closely related to awareness and preference indices than to such traditional variables as size, distance, economic, and demographic characteristics. In further analyses of the same data set, White (1981) explored the communality and content of residential preferences. First, there seemed to be substantial common agreement among the more than one thousand people in six different cities who indicated their preferences for twenty-six metropolitan areas. Second, residential preferences were more strongly related to the perceived characteristics of cities than to either their objective or informational characteristics.

In a similar vein, Jones and Zannaras (1976) investigated the variation in young adult in-migration rates for thirty Venezuelan cities. A multivariate regression model, involving traditional economic opportunity and quality-of-life variables, accounted for 65% of the variation in migration behaviour, but a second model, incorporating measures of perceived economic opportunity and perceived quality-of-life, improved the level of explanation to 83%. The authors noted that the images held by youths will not necessarily reflect those of the whole population, but point out that youths tend to form the most highly mobile segment of the population, and are therefore the most important source of potential migrants.

Jones (1978) extended this work in his discussion of so-called Venezuelan myth maps. In particular, he was able to show that these myth maps, constructed on the basis of perceived economic opportunities and quality-of-life characteristics, could be attributed to images projected by the national newspapers, which tend to popularise the regional development efforts of the public sector. Also within the context of Venezuela, the same author (Jones, 1980) has since attempted to articulate the role of cognitive variables in migration through the construction of a path model, thus isolating the direct and indirect effects of different variables.

Finally, Todd (1982) also used path analysis to explore the role of subjective attributes in explaining small-town population change in Manitoba. Given the divergence between objective and subjective correlates-of-place, he suggested that both kinds of variables could be profitably included in regression models. This suggestion was supported by the fact that, in examining small-town stability as evidenced by population change, he was able to show that the objective and subjective variables can have independent associations with in- or out-migration once the problem of multicollinerity has been identified.

Preferences

Although the explication of the links between preferences and overt behaviour has been of the utmost importance, it has been equally important to explain the preferences themselves. In other words, can we explain why certain individuals prefer certain locations? The answer to this question involves trying to establish how individuals integrate various pieces of information into some kind of overall utility value that they can then use to rank alternatives. Within the migration literature there have been three major approaches to this problem. The revealed preference approach involves the examination of observed behaviour in order to uncover the underlying preference structure, and so establish rules of behaviour. For example, Schwind (1971) inductively derived the spatial preferences of migrants for regions by analysing actual migration flows between State Economic Areas in Maine. Similarly, Tobler (1979) computed the relative attractiveness of states from data on inter-state migration. It has been pointed out, however, that only purely discretionary behaviour should be analysed in terms of revealed preferences. In those instances where the choice is constrained in some way, there will be a confounding of preferences and constraints.

Since the revealed preference approach only allows one to deduce a preferential ordering for the range of spatial alternatives that are available in a particular study area, some researchers have attempted to develop experimental designs whereby attribute values can be manipulated to produce a variety of abstract combinations, thus creating a set of hypothetical alternatives that is independent of any particular spatial structure. One of the potentially most useful approaches, within this context, is the conjoint measurement model. The conjoint measurement technique provides a method for defining a utility value for each alternative as a joint effect of its constituent attributes. The coefficients associated with those attributes indicate their individual contributions to the overall utility value.

Lieber (1979) has successfully used the conjoint measurement procedure within a migration context by asking subjects to evaluate both hypothetical and real destinations according to three predetermined attributes. The three attributes were travel time to a major city, travel time to a fresh air recreational opportunity, and travel time to close relatives. The first two attributes were split into four levels, with the third attribute being split into two levels. As a result, the experimental design consisted of thirty-two treatment combinations, and Lieber concluded that a multiplying, or nonlinear, model may be more appropriate for describing the preference judgements than a linear one.

The main advantage of the conjoint measurement model over the revealed preference approach is that subjects are directly evaluating the attributes assumed to underlie destination preferences, rather than simply the alternatives themselves. It is in this respect that the derived rules of spatial choice are considered to be independent of the particular opportunity set being considered. One of the major disadvantages of the technique, however, is that the experimental levels associated with each of the attributes have to be predetermined by the experimenter, and these levels will not necessarily coincide with the internalised thresholds for these attributes held by the subjects themselves.

A third type of methodolgy used to probe the individual decision-making process involves multidimensional scaling techniques. Such techniques can help to identify the cognitive attributes, or dimensions, that individuals use to differentiate between alternative choices. Demko (1974) sought to uncover subjects' images of a selected group of cities in Southern Ontario which were treated as potential migration destinations. The generated similarities data were subjected to multidimensional scaling analysis, in order to obtain a set of derived cognitive spaces. Similarly, Lueck (1976) used multidimensional scaling analysis to investigate the cognition of nine cities in the United States. He concluded that a three-dimensional solution was able to adequately account for the variation in the original dissimilarities data. He described these three dimensions as representing an "excitement" scale, a "cleanliness and safety" scale, and a "social milieu" scale.

As with revealed preferences and conjoint measurement, however, the use of multidimensional scaling has not been immune from criticism, especially with respect to whether it provides a useful model of the psychological processes involved in decision-making. A significant weakness is the fact that the cognitive dimensions have to be interpreted in an a posteriori fashion by the researcher, unlike conjoint measurement where the significant attributes are specified in advance. Also, the axes provided by multidimensional scaling are continuous in nature, whereas it is quite likely that individuals make decisions on the basis of certain threshold values, rather than on the basis of continuous referents (Harman and Betak, 1976).

Future work on inter-regional migration is likely to pursue a synthesis between the macro and micro-approaches (Woods, 1982). In other words, these two lines of inquiry should be regarded as being complementary rather than competitive (Golledge, 1980). Three other issues are also likely to be prominent in future work. First, temporal changes in migration patterns will be explored more explicitly through the construction of models involving lagged variables and time series analysis. Second, migration models will be restructured to incorporate a variety of individual and institutional level constraints (Lewis, 1982). It is these institutional factors, in any given migration context, that are the most amenable to governmental control, and therefore the most policy relevant (Flowerdew, 1982). Third, migration research will become increasingly sensitive to the consequences, as well as the causes, of migration (White and Woods, 1980).

INTRA-URBAN MOBILITY

A vast and growing body of literature is devoted to describing and explaining residential movement within urban areas. Almost half of the United States population made at least one residential move during the five-year period from 1970 to 1975, and 45 percent of those moves represented changes of residence within the same metropolitan area (Quigley and Weinberg, 1977). In Great Britain the figures are somewhat less, but still considerable (Short, 1978). Such high mobility rates are a major factor in shaping the socio-economic structure of urban neighbourhoods, and are often associated with the deterioration and decline of particular regions within cities.

Mobility Rates and Flows

Investigations of the spatial pattern of population turnover within cities indicate that mobility rates are much higher in the centre of the city than at the periphery. For example, Cadwallader (1982) fitted linear, power, and exponential functions to census tract data for Portland, Oregon, and found that mobility rates decreased with increasing distance from the central business district. Further exploration of the pattern was achieved through the use of trend surface analysis. For all three time periods analysed, the best fit was provided by the cubic surface, with between 21 and 57 percent of the variation in mobility rates being accounted for. Of greater interest, however, was the change over time, which suggested that the surfaces were becoming increasingly complex and convoluted.

In addition to identifying the spatial pattern of mobility rates, attempts have also been made to establish the interrelationships between mobility rates and other features of the urban environment, such as socio-economic, demographic, and housing characteristics. Using areal data for Brisbane, Australia, Moore (1969) developed a causal model involving variables like distance from the central business district, percentage of dwellings owner-occupied, and percentage of Australian born. Despite the satisfactory empirical fit, however, the model was disappointing in that the explanatory variables were selected in an ad hoc fashion, rather than on the basis of some underlying theoretical framework. More recent work in this line, has utilised the concepts of housing space and social space (Cadwallader, 1981), with the results of a series of path analyses emphasising the important role of housing type variables. Such causal systems can also be conveniently modelled using a simultaneous equations approach (Cadwallader, 1982). Such an approach explicitly recognises that mobility rates are a cause of, as well as a consequence of, neighbourhood characteristics.

Descriptions of intra-urban flows, rather than just turnover rates, have focused on the biases associated with distance and direction. As with inter-regional migration, there is a preponderence of short-distance moves. Indeed, for the city of Seattle, it has been estimated that the average length of an intra-urban move is less than three miles (Boyce, 1969). The evidence concerning directional biases is more ambiguous, however, although it has been suggested that central-area moves are randomly oriented, while moves within the suburbs tend to exhibit a sectoral bias (Clark, 1971). Such an assertion can be related to the notion that residents often possess urban images that are sectoral rather than zonal, and that the underlying socio-economic structure of cities has an important sectoral component. It is exceedingly difficult to generalise across different cities, however, as the direction of moves will be as sensitive to the idiosyncratic location of new housing development in a particular city as it is to the overall spatial pattern of cities in general.

The most significant regularity in terms of migration flows, however, is that most households move between areas of similar socio-economic status. Alperovich (1983) constructed a multiple regression model to assess the influence of several variables on migration within the Israeli city of Tel Aviv-Yafo. The estimated coefficients suggested the importance of such origin and destination characteristics as age, education, and housing quality. Besides regression analyses, entropy-maximising models also appear to have some potential in terms of predicting migration flows. Using data for Amsterdam, Clark and Avery (1978) concluded that an entropy-maximising model provides an adequate representation of the basic structure of population flows, especially given the difficulties of finding an adequate cost function for movements between small areas of a city.

The Decision to Move

Unlike the aggregate approaches described above, behavioural models of residential relocation focus on the individual decision-maker. Traditionally, the decision-making process has been conceptually partitioned into three stages; the decision to move, the search for available alternatives, and the evaluation of those alternatives (Brown and Moore, 1971). Although this compartmentalisation obviously represents an oversimplification, it has allowed researchers to focus their attention on different parts of the overall process. Studies of the initial decision to seek a new residence have emphasised the importance of previous mobility behaviour. Of particular interest in this context is the duration-of-residence effect (Clark and Huff, 1977), which suggests that the longer a household remains in a particular location the less likely it is to move.

A large number of movement decisions result from families adjusting their housing to meet the demands for space generated by changing family composition. A recent typology of reasons for moving explicates this relationship between changes in family life cycle and residential mobility, and stresses the role of housing space and tenure change (Clark and Onaka, 1983). Within the framework of path analysis, Pickvance (1974) has demonstrated that life-cycle and tenure status play a crucial role in determining both desired and expected mobility. Path analysis has also been utilised to investigate the relationships between residential satisfaction, desire to move, and actual mobility behaviour. Speare (1974) constructed an index of residential satisfaction based on housing and neighbourhood characteristics and found it to be a useful indicator of subsequent behaviour. Similarly, Hourihan (1984) developed a path model to explore how personal characteristics are causally antecedent to the perception of neighbourhood attributes in determining levels of neighbourhood satisfaction.

An alternative approach to predicting the propensity to move of individual households has revolved around the concept of residential stress (Brummell, 1981). Stress occurs because of the difference between a household's present level of satisfaction and the level of satisfaction it believes can be attained elsewhere. The amount of stress, or the magnitude of this difference, can be measured across a set of stressors, such as kind of people living in the neighbourhood and the proximity to work. Clark and Cadwallader (1973) found a reasonably high correlation between residential stress and desire to move for a group of subjects located in Santa Monica, California. Of the five stressors used in the study, size and facilities of the dwelling unit appeared to be the most important, while distance from work was the least important.

Such models can also incorporate the duration-of-residence effect by postulating a trade-off between dissatisfaction and inertia (Huff and Clark, 1978b). The level of stress, or dissatisfaction, is assumed to increase over time, as the household falls out of adjustment with its present situation. Similarly, however, with increasing duration of stay the resistance to moving, or inertia, is also expected to increase over time. Thus the probability of moving represents a trade-off between stress and inertia, where both components can be expressed as exponential functions of change over time. Despite the analytical appeal of such a conceptualisation, however, the black-box nature of the stress and inertia functions indicate that the model is best suited to a predictive rather than explanatory role (Clark, Huff and Burt, 1979). A more recent attempt to further refine the stress-inertia, or stress-resistance model has suggested that households might be as sensitive to the direction and rate of change of stress as they are to its absolute level at any specific time. Using data from the Canadian city of Saskatoon, Phipps and Carter (1984) were able to show that intention to move was greatest for those households experiencing relatively high levels of stress that had increased over the past two years. The effects of resistance on mobility were also strong, although somewhat more complex.

Residential Search

Once a household decides to move it must begin to search for alternative accommodation. There are three interrelated questions that are pivotal to any understanding of the residential search process. First, what are the information sources used to find appropriate housing? Second, how long is the search activity? Third, is there any particular spatial pattern associated with the search activity? The most important sources of information for prospective movers are newspaper advertisements, personal contacts, personal observation of "for sale" signs, and real estate agents. Different sources of information appear to be used according to the type of dwelling unit that is desired. Newspaper advertisements are often used when looking for an apartment, while real estate agents are an effective source for house hunters, especially in the context of out-of-town buyers

(Clark and Smith, 1982). Analyses of the sequential structure of information collection suggest that the preliminary stages of search are characterised by a greater dependency on newspaper advertisements, while in the later stages homebuyers become increasingly reliant on real estate agents (Clark and Smith, 1979).

The length of time spent searching depends on the degree of satisfaction associated with present alternatives, combined with the time and money costs that would be incurred by further search. In this context, a variety of probability models have been constructed to devise optimal "stopping rules" for housing search activity (Phipps and Laverty, 1983). Most search theories tend to incorporate the idea of a critical utility value which differentiates acceptable and unacceptable alternatives, but this threshold value will change during the course of searching, due to learning and preference adjustment (Smith et al., 1979). Personality factors are also likely to be involved, with conservative households being prepared to follow a satisfying rather than optimising strategy. Efforts to link the "stopping rule" models with the information acquisition approach have focused on the relationship between the costs of information derived from different sources and the length of search (Smith and Clark, 1980).

Most empirical work suggests that the spatial pattern of search is remarkably constrained. Using data from Toronto, Barrett (1976) found that more than 92% of his sample restricted themselves to average search distances of less than three miles. Although the spatial aspects of search have generally received less attention than the issues involving information sources and length of search, there are two notable expections. First, Schneider (1975) has developed a spatial search model that is especially appropriate for search involving information collected by driving around or walking around an area. Second, Huff (1982) has constructed a model in which the search procedure is described as a two-step process consisting of the initial selection of an area in which search is to be concentrated, followed by the actual selection of vacancies within that targeted area. The probability of searching within a particular area depends upon the location of the last vacancy seen by the household and the relative concentration of possible acceptable vacancies.

Neighbourhood Evaluation

The third part of the residential decision-making process, neighbourhood evaluation and choice, has two major components. First, what are the evaluative dimensions that households use to distinguish between alternative neighbourhoods? Second, how is information on those evaluative dimensions used to reach some kind of a decision? Early work on the first of these two questions tended to focus on the semantic differential technique, whereby subjects were asked to evaluate a set of neighbourhoods across a series of scales consisting of bi-polar adjectives. Principal components analysis is then used to uncover the underlying dimensions. In a study based on data collected in Christchurch, New Zealand, Johnston (1973) identified three major cognitive categories for evaluating neighbourhoods; the physical attributes, the social attributes, and the locational attributes. Similar dimensions were found by Cadwallader (1979), although the exact nature of those dimensions varied across different kinds of neighbourhoods.

An alternative approach to answering this question has involved the application of multidimensional scaling analysis. Based on data derived from a survey of women in Hamilton, Ontario, Preston (1982) concluded that residential areas are evaluated in terms of land use, lot size, social character, and housing quality. Hourihan (1979) also used multidimensional scaling analysis when examining neighbourhood perception among a sample of residents in Dublin, Ireland. The social status of the evaluated neighbourhoods turned out to be the most important differentiating characteristic, followed by familiarity and housing style. However, individual differences in the importance attached to these dimensions were apparently unrelated to the socio-economic status of the subjects. In addition, although the orthogonal principal components analyses that are generally utilised in semantic differential investigations assume the evaluative dimensions are independent, multidimensional scaling approaches suggest that they might be highly related.

Residential Choice

Attempts to model the actual choice process involved in residential mobility have included using the conjoint measurement technique and information integration theory. The conjoint measurement model requires subjects to evaluate multi-attribute alternatives by comparing predetermined levels of the supposedly important attributes. The subjective rankings of various combinations of these experimental levels are then decomposed into preference functions. For example, Knight and Menchik (1976) have explored individual preferences for a variety of residential forms based on the ranking of hypothetical levels of such attributes as distance between houses, view from the backyard, and price. Information integration theory involves constructing an algebraic model of human information-processing, and then testing this model by means of analysis of variance. The theory is concerned with how different pieces of information are integrated into a single overall evaluation, and in a study of student judgements as to the residential desirability of various neighbourhoods,

Louviere and Meyer (1976) provided evidence for a linear averaging kind of combination rule.

An increasingly popular approach for analysing discrete choice situations involves using log-linear models. Such models represent a group of procedures for investigating the associations among a set of categorical variables. Discrete multivariate analysis allows a large number of parameters to be estimated, including ones for each of the category levels, as well as any important interactions among them. It should be noted, however, that models which contain third-order interactions, or more than three second-order interactions, are often very difficult to interpret. The most that can be said in such situations is that the association structure is extremely complex. Whitney and Boots (1978) have used a long-linear modelling approach to assess the impact of selected household characteristics on residential mobility in the twin cities of Kitchener and Waterloo, Ontario. In a similar vein, Segal (1979) has provided an example of log-linear modelling in the context of neighbourhood choice.

It is clear that the process of residential choice requires households to choose between a number of alternatives that are each described by many attributes. As such we are dealing with a complex process, and most choice models make the perhaps unwarranted assumption that individuals can consider all the attributes describing each alternative before coming to a decision. The elimination-by-aspects model, however, circumvents this problem by explicitly postulating a sequential process in which individuals can eliminate a large number of alternatives after only a few attributes have been considered. The decision-maker selects one aspect, or attribute, and then eliminates all alternatives that are unsatisfactory with respect to that single attribute. Then a second attribute is chosen, and so on, until only one alternative remains. In this way the decision-maker is not expected to integrate all the pertinent information before arriving at one ultimate decision, but rather, the problem is decomposed into a series of decision steps.

Within the context of residential mobility we can hypothesise at least two such steps. First, the household might collect information to help select an appropriate neighbourhood. Second, information is obtained to choose a specific house within that neighbourhood. Using data from Syracuse, New York, Talarchek (1982) found that behavioural patterns of residential search and selection are highly individual, but that the sequence in which information is acquired does indicate a general two-stage process of the kind just described. When investigating the locational choice process of some new residents in Melbourne, Australia, Young (1984) also found that the elimination-by-aspects model provided an acceptable statistical fit to the data.

It should be remembered, however, when modelling residential choice, that one should never expect a perfect correlation between preferences and behaviour, as all decisions are acted out within a series of contextual constraints. The constraints that operate within the context of residential mobility involve obstacles that produce attitude-discrepant behaviour by restricting the opportunity set, influencing the formation of preferences, or preventing choice actualisation (Desbarats, 1983). These constraints can be either individual or institutional in their origin. At the individual level a potential migrant's income will obviously restrict where he or she can live, regardless of the underlying preferences. At the institutional level, real estate agents have a significant impact on migration flows because they are a major repository of information concerning housing vacancies. Similarly, financial institutions such as banks and savings and loan associations influence mobility patterns through their regulation of the flow of mortgage money into the housing market.

Future research on residential mobility is likely to be more balanced with respect to the consideration of demand and supply side characteristics. Existing models of the residential decision-making process neglect the role of various institutional constraints, thus implying a situation of consumer sovereignty. Greater attention to the behaviour of supply side actors will not only increase the predictive capacities of such models, but will also facilitate the development of governmental intervention strategies for improving housing consumption (Moore, 1982). These links between mobility and public policy will be increasingly scrutinised as governments attempt to understand the impacts of their neighbourhood programmes.

Acknowledgements

The study was supported by grants from the Wisconsin Alumni Research Foundation and the National Science Foundation, grant number SES-8206940. Part of the work was undertaken while the author was a Visiting Scholar in the Department of Geography, University of Cambridge.

REFERENCES

Alperovich, G. (1983) 'Lagged Response in Intra-Urban Migration of Home Owners', Regional Studies, 17, 297-304

Barrett, F. (1976) 'The Search Process in Residential Relocation', Environment and Behavior, 8, 169-198

Berry, B.J.L. (1966) Essays on Commodity Flows and the Spatial Structure of the Indian Economy, Department of

Geography Research Paper No. 111, University of Chicago
Blaug, M. (1976) 'The Empirical Status of Human Capital Theory: A Slightly Jaundiced View', Journal of Economic Literature, 14, 827-855
Boyce, R.R. (1969) 'Residential Mobility and its Implications for Urban Spatial Change', Proceedings of the Association of American Geographers, 1, 22-26
Brown, L.A. and Moore, E.G. (1971) 'The Intra-Urban Migration Process: A Perspective' in L.S. Bourne (ed.), Internal Structure of the City, Oxford University Press, New York, pp. 200-209
Brummell, A.C. (1981) 'A Method of Measuring Residential Stress', Geographical Analysis, 13, 248-261
Cadwallader, M. (1978) 'Urban Information and Preference Surfaces: Their Patterns, Structures, and Interrelationships', Geografiska Annaler, 60B, 97-106
Cadwallader, M. (1979) 'Neighborhood Evaluation in Residential Mobility', Environment and Planning, A11, 393-401
Cadwallader, M. (1981) 'A Unified Model of Urban Housing Patterns, Social Patterns, and Residential Mobility', Urban Geography, 2, 115-130
Cadwallader, M. (1982) 'Urban Residential Mobility: A Simultaneous Equations Approach', Transactions of the Institute of British Geographers, New Series 7, pp.458-473
Cadwallader, M. (1985) 'Structural-Equation Models of Migration: An Example from the Upper Midwest USA', Environment and Planning, A17, 101-113
Cebula, R.J. (1980) 'Voting with One's Feet: A Critique of the Evidence', Regional Science and Urban Economics, 10, 91-107
Clark, G.L. (1982a) 'Dynamics of Interstate Labor Migration', Annals of the Association of American Geographers, 72, 297-313
Clark, G.L. (1982b) 'Volatility in the Geographical Structure of U.S. Interstate Migration', Environment and Planning, A14, 145-167
Clark, W.A.V. (1971) 'A Test of Directional Bias in Residential Mobility' in H. McConnel and D.W. Yaseen (eds.), Models of Spatial Variation, Northern Illinois University Press, De Kalb, Illinois, pp. 2-27
Clark, W.A.V. (1982) 'Recent Research on Migration and Mobility: A Review and Interpretation', Progress in Planning, 18, 1-56
Clark, W.A.V. and Avery, K.L. (1978) 'Patterns of Migration: A Macroanalytic Case Study' in D.T. Herbert and R. J. Johnston (eds.), Geography and the Urban Environment, Volume I, John Wiley, Chichester, pp.135-196

Clark, W.A.V. and Cadwallader, M. (1973) 'Locational Stress and Residential Mobility', Environment and Behavior, 5, 29-41

Clark, W.A.V. and Huff, J.O. (1977) 'Some Empirical Tests of Duration-of-Stay Effects in Intraurban Migration', Environment and Planning, A9, 1357-1374

Clark, W.A.V. and Onaka, J.L. (1983) 'Life Cycle and Housing Adjustment as Explanations of Residential Mobility', Urban Studies, 20, 47-57

Clark, W.A.V. and Smith, T.R. (1979) 'Modeling Information Use in a Spatial Context', Annals of the Association of American Geographers, 69, 575-588

Clark, W.A.V. and Smith, T.R. (1982) 'Housing Market Search Behavior and Expected Utility Theory: 2. The Process of Search', Environment and Planning, A14, 717-737

Clark, W.A.V., Huff, J.O. and Burt, J.E. (1979) 'Calibrating a Model of the Decision to Move', Environment and Planning, A11, 689-704

Clayton, C. (1977) 'Interstate Population Migration Process and Structure in the United States, 1935-70', Professional Geographer, 29, 177-181

Dahlberg, A. and Holmlund, B. (1978) 'The Interaction of Migration, Income, and Employment in Sweden', Demography, 15, 259-266

Demko, D. (1974) 'Cognition of Southern Ontario Cities in a Potential Migration Context', Economic Geography, 50, 20-34

Desbarats, J.M. (1983) 'Spatial Choice and Constraints on Behavior', Annals of the Association of American Geographers, 73, 340-357

Dorigo, G. and Tobler, W. (1983) 'Push-Pull Migration Laws', Annals of the Association of American Geographers, 73, 1-17

Flowerdew, R. (1982) 'Institutional Effects of Internal Migration' in R. Flowerdew (ed.), Institutions and Geographical Patterns, St. Martin's Press, New York, pp.209-227

Fotheringham, A.S. (1981) 'Spatial Structure and Distance-Decay Parameters', Annals of the Association of American Geographers, 71, 425-436

Gober-Myers, P. (1978) 'Interstate Migration and Economic Growth: A Simultaneous Equations Approach', Environment and Planning, A10, 1241-1252

Golledge, R.G. (1980) 'A Behavioral View of Mobility and Migration Research', Professional Geographer, 32, 14-21

Gould, P.R. and White, R. (1974) Mental Maps, Penguin, London

Grant, K.E. and Vanderkamp, J. (1980) 'The Effects of Migration on Income: A Macro Study with Canadian Data, 1965-71', Canadian Journal of Economics, 13, 381-406

Greenwood, M.J. (1970) 'Lagged Response in the Decision to Migrate', Journal of Regional Science, 10, 375-384

Greenwood, M.J. (1975) 'Research on Internal Migration in the United States: A Survey', Journal of Economic Literature, 13, 397-433

Greenwood, M.J. (1981) Migration and Economic Growth in the United States: National, Regional, and Metropolitan Perspectives, Academic Press, New York

Grigg, D.B. (1977) "E.G. Ravenstein and the 'Laws of Migration'", Journal of Historical Geography, 3, 41-54

Hagerstrand, T. (1957) 'Migration and area' in D. Hannerberg et al. (eds.), Migration in Sweden, Gleerup, Lund, pp.27-158

Harman, E. and Betak, J. (1976) 'Behavioral Geography, Multidimensional Scaling, and the Mind' in R.G. Golledge and G. Rushton (eds.), Spatial Choice and Spatial Behavior, Ohio State University Press, Columbus, Ohio, pp.3-20

Harris, J.R. and Todaro, M.P. (1970) 'Migration, Unemployment and Development: A Two Sector Model', American Economic Review, 60, 126-142

Haynes, R.M. (1974) 'Application of Exponential Distance Decay to Human and Animal Activities', Geografiska Annaler, 56B, 90-104

Hourihan, K. (1979) 'The Evaluation of Urban Neighborhoods: 1. Perception', Environment and Planning, A11, 1337-1353

Hourihan, K. (1984) 'Residential Satisfaction, Neighbourhood Attributes, and Personal Characteristics: An Exploratory Path Analysis in Cork, Ireland', Environment and Planning, A16, 425-436

Huff, J.O. (1982) 'Spatial Aspects of Residential Search' in W.A.V. Clark (ed.), Modelling Housing Market Search, Croom Helm, London, pp. 106-129

Huff, J.O. and Clark, W.A.V. (1978a) 'The Role of Stationarity in Markov and Opportunity Models of Intraurban Migration' in W.A.V. Clark and E.G. Moore (eds.), Population Mobility and Residential Change, Studies in Geography No. 25, Northwestern University, Illinois, pp. 183-213

Huff, J.O. and Clark, W.A.V. (1978b) 'Cumulative Stress and Cumulative Inertia: A Behavioral Model of the Decision to Move', Environment and Planning, A10, 1101-1119

Johnston, R.J. (1973) 'Spatial Patterns in Suburban Evaluations', Environment and Planning, 5, 385-395

Jones, R. (1978) 'Myth Maps and Migration in Venezuela', Economic Geography, 54, 75-91

Jones, R. (1980) 'The Role of Perception in Urban In-Migration: A Path Analytic Model', Geographical Analysis, 12, 98-108

Jones, R. and Zannaras, G. (1976) 'Perceived Versus Objective Urban Opportunities and the Migration of Venezuelan Youths', Annals of Regional Science, 10, 83-97
Kau, J. and Sirmans, C. (1977) 'The Influence of Information Cost and Uncertainty on Migration: A Comparison of Migrant Types', Journal of Regional Science, 17, 89-96
Kleiner, M. and McWilliams, W. (1977) 'An Analysis of Alternative Labor Force Population Forecasting Models', Annals of Regional Science, 11, 74-85
Knight, R.L. and Menchik, M.D. (1976) 'Conjoint Preference Estimation for Residential Land Use Policy Evaluation' in R.G. Golledge and G. Rushton (eds.), Spatial Choice and Spatial Behavior, Ohio State University Press, Columbus, Ohio, pp. 135-155
Lewis, G.J. (1982) Human Migration: A Geographical Perspective, St. Martin's Press, New York
Lieber, S.R. (1978) 'Place Utility and Migration', Geografiska Annaler, 60B, 16-27
Lieber, S.R. (1979) 'An Experimental Approach for the Migration Decision Process', Tijdschrift voor Economische en Sociale Geografie, 70, 75-85
Lloyd, R. (1976) 'Cognition, Preference, and Behavior in Space: An Examination of the Structural Linkages', Economic Geography, 52, 241-253
Louviere, J.J. and Meyer, R.J. (1976) 'A Model for Residential Impression Formation', Geographical Analysis, 8, 479-486
Lowry, I.S. (1966) Migration and Metropolitan Growth: Two Analytical Models, Chandler, San Francisco
Lueck, V.M. (1976) 'Cognitive and Affective Components of Residential Preferences for Cities: A Pilot Study' in R.G. Golledge and G. Rushton (eds.), Spatial Choice and Spatial Behavior, Ohio State University Press, Columbus, Ohio, pp. 273-300
Moore, E.G. (1969) 'The Structure of Intra-Urban Movement Rates: An Ecological Model', Urban Studies, 6, 17-33
Moore, E.G. (1982) 'Search Behavior and Public Policy: The Conflict Between Supply and Demand Perspectives' in W.A.V. Clark (ed.), Modelling Housing Market Search, Croom Helm, London, pp. 224-238
Phipps, A.G. and Carter, J.E. (1984) 'An Individual-Level Analysis of the Stress-Resistance Model of Household Mobility', Geographical Analysis, 16, 176-189
Phipps, A.G. and Laverty, W.H. (1983) 'Optimal Stopping and Residential Search Behavior', Geographical Analysis, 15, 187-204
Pickles, A. and Rogerson, P. (1984) 'Wage Distributions and Spatial Preferences in Competitive Job Search and Migration', Regional Studies, 18, 131-142

Pickvance, C.G. (1974) 'Life Cycle, Housing Tenure and Residential Mobility: A Path Analytic Approach', Urban Studies, 11, 171-188
Preston, V. (1982) 'A Multidemensional Scaling Analysis of Individual Differences in Residential Area Evaluation', Geografiska Annaler, 64B, 17-26
Quigley, J.M. and Weinberg, D.H. (1977) 'Intra-Metropolitan Residential Mobility: A Review and Synthesis', International Regional Science Review, 2, 41-66
Rees, P.H. and Wilson, A.G. (1977) Spatial Population Analysis, Edward Arnold, London
Riddell, J.B. (1970) 'On Structuring a Migration Model', Geographical Analysis, 2, 403-409
Rodgers, A. (1970) 'Migration and Industrial Development: The Southern Italian Experience', Economic Geography, 46, 111-135
Rodgers, A. (1967) 'A Regression Analysis of Interregional Migration in California', The Review of Economics and Statistics, 49, 262-267
Rogerson, P.A. (1984) 'New Directions in the Modelling of Interregional Migration', Economic Geography, 60, 111-121
Rummel, R.J. (1965) 'A Field Theory of Social Action with Application to Conflict within Nations', General Systems, 10, 183-204
Rushton, G. (1981) 'The Scaling of Locational Preferences' in K. Cox and R. Golledge (eds.), Behavioral Problems in Geography Revisited, Methuen, New York, pp. 67-92
Schneider, C.H. (1975 'Models of Space Searching in Urban Areas', Geographical Analysis, 7, 173-185
Schwind, P.J. (1971) 'Spatial Preferences of Migrants for Regions: The Example of Maine', Proceedings of the Association of American Geographers, 3, 150-156
Schwind, P.J. (1975) 'A General Field Theory of Migration: United States, 1955-1960', Economic Geography, 51, 1-16
Segal, D. (1979) 'A Quasi-Loglinear Model of Neighborhood Choice' in D. Segal (ed.), The Economics of Neighborhood, Academic Press, New York, pp. 57-82
Short, J.R. (1978) 'Residential Mobility', Progress in Human Geography, 2, 419-447
Slater, P.B. (1984) 'A Partial Hierarchical Regionalization of 3140 U.S. Counties on the Basis of 1965-1970 Intercounty Migration', Environment and Planning, A16, 545-550
Smith, T. and Clark, W.A.V. (1980) 'Housing Market Search: Information Constraints and Efficiency' in W.A.V. Clark and E.G. Moore (eds.), Residential Mobility and Public Policy, Sage Publications, Beverly Hills
Smith, T., Clark, W.A.V., Huff, J.O. and Shapiro, P. (1979) 'A Decision-Making and Search Model for Intra-urban Migration', Geographical Analysis, 11, 1-22

Speare Jr., A. (1971) 'A Cost-Benefit Model of Rural to Urban Migration in Taiwan', Population Studies, 25, 117-130

Speare Jr., A. (1974) 'Residential Satisfaction as an Intervening Variable in Residential Mobility', Demography, 11, 173-188

Talarchek, G.M. (1982) 'Sequential Aspects of Residential Search and Selection', Urban Geography, 3, 34-57

Tobler, W.R. (1979) 'Estimation of Attractivities from Interactions', Environment and Planning, A11, 121-127

Todd, D. (1982) 'Subjective Correlates of Small-Town Population Change', Tijdschrift voor Economische en Sociale Geografie, 73, 109-121

White, P. and Woods, R. (eds.) (1980) The Geographical Impact of Migration, Longman, London

White, S. (1974) 'Residential Preference and Urban In-Migration', Proceedings of the Association of American Geographers, 6, 47-50

White, S. (1978) 'Mental Map Variability: A Migration Modeling Problem', Annals of Regional Science, 12, 89-97

White, S. (1980) 'Awareness, Preference, and Interurban Migration', Regional Science Perspectives, 10, 71-86

White, S. (1981) 'The Influence of Urban Residential Preferences on Spatial Behavior', Geographical Review, 71, 176-187

Whitney, J.B. and Boots, B.N. (1979) 'An Examination of Residential Mobility through the Use of the Log-Linear Model', Regional Science and Urban Economics, 9, 393-409

Wolpert, J. (1965) 'Behavioral Aspects of the Decision to Migrate', Papers and Proceedings of the Regional Science Association, 15, 159-169

Wolpert, J. (1967) 'Distance and Directional Bias in Inter-Urban Migratory Streams', Annals of the Association of American Geographers, 57, 605-616

Woods, R. (1982) Theoretical Population Geography, Longman, London

Young, W. (1984) 'Modelling Residential Location Choice', Australian Geographer, 16, 21-28

Chapter Ten

POPULATION MODELLING

P. Rees

ON MODELS AND POPULATION

Social scientists look at the world from a large number of viewpoints - methodological, political, philosophical. Very often they attempt to make sense of what they observe by simplifying their view into a set of propositions about human behaviour. These propositions may be set out in forms that can be verified or rejected by comparison with reality (positivist social science) or in forms that do not depend on such comparison (dialectical social science). In this chapter attention will be concentrated on representations of reality or models that fall between the positivist and dialectical views. The models discussed are quantitative statements about how populations develop that derive from logical argument about the processes that affect those populations, which can be filled with empirical content, but for which tests of empirical validity are rarely formulated.

Two concerns have motivated population modellers: the first is a desire to understand the behaviour of populations in the past; the second is a concern to understand how populations may develop in the future. The models discussed in this chapter concentrate on the latter concern.

A wide variety of different populations have been modelled at many spatial scales. The commonest unit of analysis has been the individual aggregated for some set of spatial units. Groups of individuals such as households have also been used, though less often. Occasionally, the spatial unit itself has been used and its evolution between population states over time examined. The difficulty of analysis rises as one proceeds from modelling populations of individuals to modelling populations of households to modelling populations of areas: the consensus of experience is that, if possible, it is best to work out the nature of households and the character of areas from the actions of individuals.

What is attempted in this chapter is an account of some of the issues which researchers seeking to build explicit quantitative models of population dynamics need to keep in mind. The examples are drawn mainly from the author's experience with multi-regional population models but many of the issues apply equally to other families of models such as those based on microsimulation principles (Clarke, Keys and Williams, 1981) or spatial interaction principles (Ledent, 1985; Stillwell, 1983; Willekens and Baydar, 1985).

In the next section of the chapter some fundamental concepts underlying population modelling are described, initially in relation to an ideal data set. The following section then constructs population projection models using those fundamental concepts. The final section of the chapter introduces some simplifications of the matrix models and complications, that is, aspects of population dynamics not included in the earlier review of concepts and models.

FUNDAMENTAL CONCEPTS

An Ideal Data Set and a Life History Framework

An ideal data set for investigating population dynamics would consist of the life histories of a population in which all the attributes of an individual and the changes in those attributes that take place are recorded continuously. Figure 10.1 shows an abstract representation of such a set. The system of interest is bounded, both spatially and temporally; the system is divided into 4 states; the system has a population stock at the start (persons B, C in the diagram) and a population stock at the end; the system receives entries of individuals from the outside world over time (persons A, D in the diagram) and loses members to the outside world (persons C, D).

An example that begins to approximate to this ideal data set is the OPCS Longitudinal Study (Brown and Fox, 1984) in which the Office of Population Censuses and Surveys (OPCS) has added vital events records over the period 1971-81 to a 4 birthdays in 365 (just over 1%) sample of the population usually resident in England and Wales at the 1971 census, and is currently linking those records with the entries for the same persons traceable in the 1981 census. The focus to date in studies using this data set has been the relationships between changes of state in the 1971-81 period (such as mortality, cancer incidence or migration) to characteristics recorded in the 1971 census at the start of the period (such as housing tenure, social class or ethnicity).

The view of the outside world taken in Figure 10.1 has also been used in projective studies (Stone, 1971, 1975) but for most purposes it is necessary to decompose entries and exits into categories that more closely match real events. Entrants to the system can be divided into births (new persons) and immigrants (persons entering from the outside

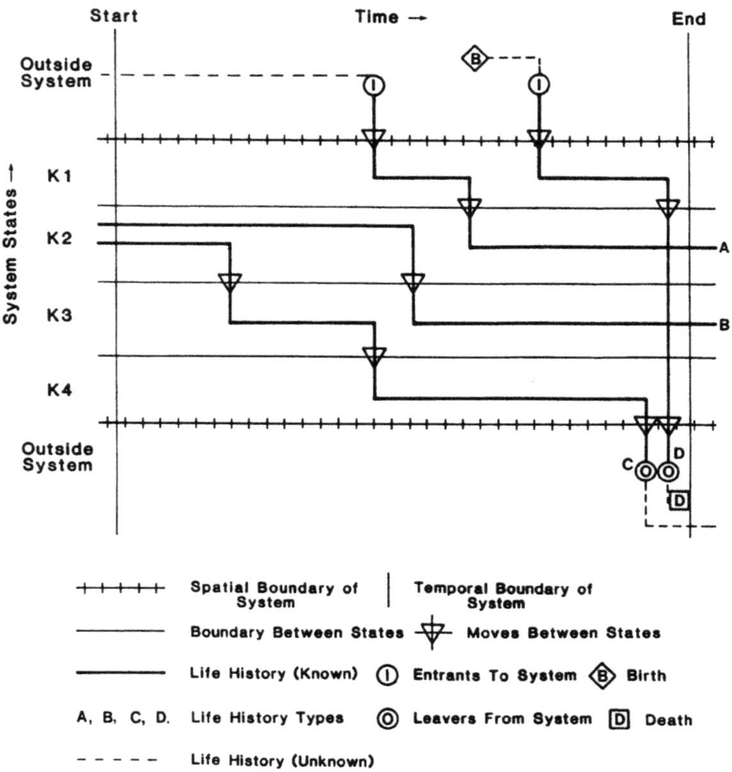

Fig. 10.1 Life histories in an ideal data set.

world); leavers can be broken down into deaths (persons dying) and emigrants (persons exiting to the outside world). A full decomposition recognises four starting life states for persons and four ending life states (Rees and Wilson, 1973; Rees, 1984a), yielding sixteen possible kinds of life history. Such a scheme is shown in Table 10.1. Or, because the detailed information for such classifications is often not available, a condensed version with three starting life states is often useful (Table 10.2).

The importance of such schemes is that they enable the researcher to recognise clearly the position of his or her data set with respect to an ideal framework, and, in particular, to work out what subpopulations have been missed and what biases are present in the data set being employed. For example, in Table 10.3 the information given in Brown and

Table 10.1: Sixteen Life History Types

Initial states			Final states: Survival at end, Inside system si	Survival at end, Outside system so	Death in interval, Inside di	Death in interval, Outside do	Totals
Exist at start	Inside system	ei	a	b	c	d	Starting population
	Outside system	eo	e	f	g	h	Existing immigrants
Birth in period	Inside system	bi	i	j	k	l	Births
	Outside system	bo	m	n	o	p	Infant immigrants
Totals			End population	Surviving emigrants	Deaths	Non-surviving emigrants	All persons

life history quadrant

Table 10.2: Nine Life History Types

Initial states			Final states: Inside system: Survival at end si	Final states: Inside system: Death in interval di	Final states: Outside system .o	Totals
Inside system	Existence at start	ei	q	r	s	Starting population
	Birth in interval	bi	t	u	v	Births in system
Outside system		.o	w	x	y	Existing immigrants
Totals			Ending population	Deaths in system	Emigrants	All persons

Fox (1984) on the numbers of people captured in the Longitudinal Study who experience various events has been re-assembled and re-estimated in a three initial state-three final state framework. Initially, 530,000 sample members were selected from the 1971 census, but only 513,000 could be traced in the National Health Service Central Registrar (NHSCR). Some 71,000 births and 28,000 immigrants were added to the sample over the decade from the four birthdays and 60,000 deaths occurred to sample members. Emigrants surrendering their NHS cards number 14,000 but this was clearly an undercount as OPCS estimate that there was a slight emigration loss to the England and Wales population over the 1971-81 decade. If the number of emigrants is estimated as 30,000 then the remaining numbers in the table (Table 10.3) follow from the accounting arithmetic, using marginal distributions where necessary. Of the 429,000 sample members who are estimated to have appeared in both 1971 and 1981 censuses, only 405,000 could be found, of whom 400,000 could also currently be traced in the NHSCR. It is these 400,000 with histories linked between 1971 and 1981 who are in the process of being intensively studied. This will be appropriate for most purposes, but researchers will need to bear in mind that this data set contains only 65% of the life histories originally sampled.

The accounting framework used here to analyse the nature of life history data sets has been employed in regional population projections based on aggregate data, but with one slight difference. The transitions of persons from outside the system to outside the system again (for example, life history D in Figure 10.1; cells f, h, m and p in Table 10.1; cell y in Table 10.2) have generally been ignored as they contribute to neither the initial population stocks nor to the final. However, if the focus of interest is on behavioural relationships in a population, cognisance must be taken of these persons, estimated as 1000 in the Longitudinal Study sample, as part of their lives are spent within the system of interest.

States of the System

Discussion to date has concerned the macro-states or life-states of the population denoted in Tables 10.1 and 10.2 using letters such as "si", survival inside the system at the end of the time interval. However, of more intrinsic interest will be the ways in which the population is described or classified, and the movement of individuals between such classes over their lifetimes.

Set out in Figure 10.2 are examples of the kind of state to state transfers that have been built into population models. Transfers allowed by the logic of the classifications have been shaded.

Table 10.3: An Estimate of the Number of Persons in the OPCS Longitudinal Study, 1971-81, in a Classification Using Nine Life History Types

Initial states (all populations are in 1000s)		Final states: England and Wales: Survival Census 1981	Final states: England and Wales: Death 1971-81	Final states: Outside England and Wales	Totals
England and Wales	Census 1971	(429)	(56)	(28)	513
	Birth 1971-81	(67)	(3)	(1)	71
Outside England and Wales		26	(1)	(1)	28
Totals		(522)	60	(30)	612

Source: Figures assembled from Brown and Fox (1984). Those in brackets have been estimated (e.g. emigrants).

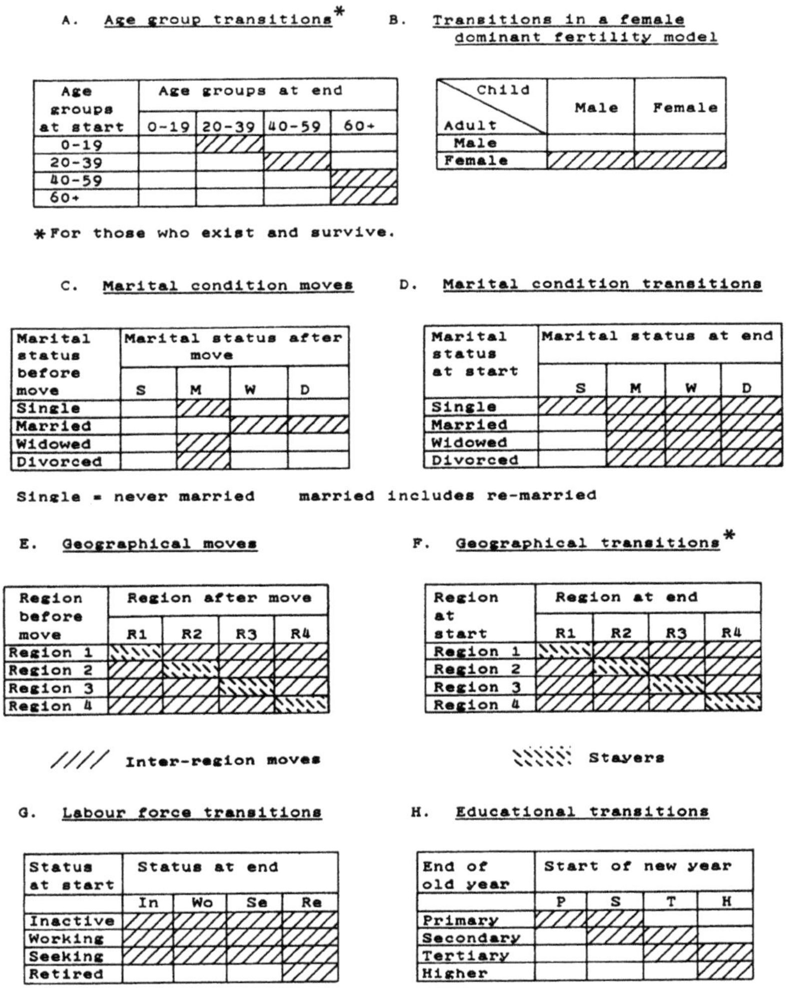

Fig. 10.2 Examples of state to state transfers employed in population models.

Most frequently incorporated are age group to age group transfers (Figure 10.2A), which in most cases involve transfer from one age group to the next highest. If the population model design allows any other transfer (except in the last open ended age group) it must be treated sceptically. The exact meaning of the "age group" will differ depending on which life history quadrant the transitions fall in, and what age-time observation plan is used.

Normally, in population models the populations of both sexes are used because their observed behaviour differs. Transfers between the sexes occur in only one respect in the fertility component of a population projection model: infants of either sex are produced by multiplying female populations at risk by the appropriate fertility rates. The marriage process is usually represented in demographic models by moves by the individual from an unmarried to a married state (Figure 10.2C). The instanteous moves allowed between marital statuses are logically restricted by their definition: no-one who has been married can become never-married; a divorced person cannot become widowed directly, a marriage must intervene. However, if we compare a person's marital condition at one point in time with that at another, then transitions between most marital statuses can take place (Figure 10.2D), particularly if the time interval is lengthy.

This distinction between instanteous moves and transitions over a time interval is an important one in the design of a population model incorporating migration between geographical places. Although the migration matrices in Figures 10.2E and 10.2F are similarly shaded, they are distinguished conceptually and should be used with different population projection models (see Rees, 1984a).

Labour force transitions, as represented in Figure 10.2G, are fairly straightforward in principle, though measurement of all transitions is difficult because the inactive and retired tend to escape the statistical nets that capture the employed and those seeking work (the unemployed).

The example in Figure 10.2H of transitions in the educational system illustrates the care that must be taken in definitions of states. Whereas most transfers take place over the course of a year, in schools and colleges transfers occur between the end of one educational year and the beginning of the next. This can, for example, cause confusion when models in the educational sector are compared with those in other sectors.

Most population models will involve combinations of the characteristics displayed in Figure 10.2. In some cases this is essential if reasonable results are to be obtained. For example, Figure 10.2H's educational transitions must be disaggregated carefully by age: otherwise if age-aggregated transitions probabilities were to be used in the usual multiplicative models, then pupils would be seen surviving in grades longer than the rules of the educational system allowed. The most usual combination involves age, sex and one other attribute of interest. Expanding the state space of the model further than these three dimensions quickly leads to difficulties of too sparse matrices of transitions. For instance, if the fairly crude classifications of Figure 10.2 were combined, the number of states into which the population had to be divided would be 2048 with over 150,000 state to state transitions to be incorporated into the population model. A brief review of alternative strategies for dealing with this combinatorial problem is given below.

Type of Transfer

There was considerable confusion in the 1970s, when the conventional cohort-survival and life expectancy models were generalised to incorporate an additional set of states such as geographical regions or marital statuses, as to the relationship between type of transfer data and appropriate model (see Rees, 1983a for detailed sources and discussion). What was at issue was not the form of the multistate cohort- survival or life expectancy model, but rather the way in which the multistate transition probabilities should be calculated from different kinds of transfer data, and how the interpretation of model outputs differed depending on transfer data type.

Transfers can be measured in one of three ways: either as transitions or as moves or as last migrations. A transition is a change of a person's state from that at the start of a time interval to that at the end. A move is a change between a prior state to a following state during a time interval. A last migration is the change between the current state at the end of a time interval and the last previous state that was different, within the time interval of measurement. If we have an ideal life history data set, transfers can be measured in any of these ways. Consider the life histories in Figure 10.1. Between states K2 and K3 there are two moves (by persons B and C) but only one transition (person B) and one last migration or migrant (person B). Between states K1 and K2 there are two moves (by persons A and D), one last migration (person A) but no transitions because person A would be counted in the outside system-state K2 transition, not that from K1 to K2. In general, there are more moves than last migrations than transitions (Rees, 1984a). The number of transitions and last migrations can never exceed the number of persons in the system, but there is no such restriction on the number of moves. Using both movement data and transition data it is possible to link opening and closing stocks of a population and thus to develop population projection models (Willekens and Drewe, 1984; Rees, 1984b). Life tables can be constructed from both movement and transition perspectives (Ledent, 1980, Ledent and Rees, 1980). It is not possible, however, to employ last migration data in such models directly (Courgeau, 1980) until they have been converted into transition data estimates (Rees, 1984a).

Therefore, in developing a multistate model of a population of interest it is necessary to first ascertain what type of transfer data are available and then to adopt the appropriate model. Thus, Willekens and Drewe (1984), in their projection model for the provinces of the Netherlands, use Register derived migration data which constitute moves rather than transitions, and therefore use a movement based projection model. Ram and Rees (1985) have available migration data of a transition nature from a retrospective sample survey in their study of the Indian population of Bradford, and therefore adopt a transition based projection model.

In both these studies use was made of demographic accounts, which are consistent tabulations of the data input to the base period of the projection model and an invaluable aid to the proper estimation of the necessary transition or movement rates. Table 10.4 sets out transition accounts and movement accounts tables for one age transition.

In the transition accounts table (Table 10.4A), the same variable, K_a, is used throughout to represent persons in an age transition classified by row and column states shown in the superscript list whereas different letters are used in the movement accounts table to represent different events. The superscript list consists of a macrostate label for initial life state together with the associated mesostate location label followed by a macrostate label for final life state together with its associated location (or other state) label. For example, K_a^{e2s1} represents the number of persons in an age transition a existing in region 2 at the start of the time interval who migrate to and survive in region 1 at the end of the interval. Such information is supplied in many census migration tables based on a retrospective question about where people were resident one or five years before the census. Careful estimation is required of the data that are input to population accounts and a modelling system is used to fill out the table in a consistent fashion (see Rees, 1981 for full details of operations software).

The example of transition accounts for South East England (Table 10.4B) shows how the populations of the South East and the Rest of Great Britain of 1.184 and 2.385 millions transfer over the 1966-71 quinquennium. Some 16% of the South East's 20-24 year olds migrate out of the region, to be only partially replaced by in-migrants. Only 9% of the Rest of Britain's population out-migrates but losses are higher absolutely.

Movement accounts (Table 10.4C) have a simpler structure in that the regional deaths (D_a^1, D_a^2) are entered in the rows in which the deceased were living prior to their final move rather than as column totals ($K_a^{e.d1}$, $K_a^{e.d2}$) to existence-death transition matrix (which has to be estimated rather than measured directly). The counts of interregional moves are entered in the same positions as the corresponding counts of exist-survive transitions. The two types of accounts differ in the nature of the diagonal terms: the K_a^{eisi} terms in transition accounts represent counts of surviving stayers who may be counted directly in the census tables or worked out as residuals by subtracting from the initial or final populations the other elements in the corresponding row or column. The diagonal terms in movement accounts are, however, always worked out as residuals and merely serve as arithmetic balancing terms.

Although it is not possible to compare the example of movement accounts directly because the spatial systems differ, the level of movements is clearly much higher than the level of transitions (Table 10.4B): this is unlikely to represent a real difference in mobility levels between the two periods. In general, the larger the time interval used in measuring demographic transfers the greater will be the differences between transition, movement and last migrant measures of multistate mobility. Conversely, the smaller the interval the closer is the agreement between the measures. At the limit of instanteous observation there is only one measure of intrinsic mobility.

Age-Time Observation Plan

The final set of concepts which need to be thoroughly understood by the population modeller concern the age-time plan (ATP) used in observing or measuring demographic variables. There are four ATPs usually involved and their nature is best explained graphically (Figure 10.3) using four Lexis diagrams (Lexis was the German demographer who invented the diagram).

The diagrams have age as their vertical axis and time as their horizontal. They also have, in effect, a third dimension along the principal diagonal, that of cohort. Persons born in the same time interval move together diagonally (at 45 degrees if the age and time scales are the same) through the diagram as they age with the passage of time. Projection models, whether concerned with the evolution of population sizes or the duration of the lives of the population, work with cohorts either observed in a period between two age groups (ATP2 in Figure 10.3) or observed over two periods between two birthdays (ATP3 in Figure 10.3).

However, ATP2 is only rarely and ATP3 is hardly ever used to observe the demographic inputs to population models. One data set which is collected in ATP2 form is that of transitions in census migration tables. These tables report migration by age group at the end of the time interval and the initial age group must be inferred. Sometimes this leads to a failure to measure migration for the first cohort, those born during the time interval. In many countries (but not the UK) care is taken to employ the date of birth of individuals experiencing events to doubly classify those events by age at time of the event and by cohort of birth. This means that data are observed using ATP4 (Figure 10.3) within Lexis dia-

Table 10.4: Population Accounts: Structure and Examples

A. Transition accounts: algebraic variables

Initial states	Final states: Survival at end in R1	R2	RW	Death in interval in R1	R2	RW	Totals
Region 1	K_a^{e1s1}	K_a^{e1s2}	K_a^{e1so}	K_a^{e1d1}	K_a^{e1d2}	K_a^{e1so}	$K_a^{e1..}$
Region 2	K_a^{e2s1}	K_a^{e2s2}	K_a^{e2so}	K_a^{e2d1}	K_a^{e2d2}	K_a^{e2do}	$K_a^{e2..}$
Rest World	K_a^{eos1}	K_a^{eos2}	0	K_a^{eod1}	K_a^{eod2}	0	$K_a^{eo..}$
Totals	$K_a^{e.s1}$	$K_a^{e.s2}$	$K_a^{e.so}$	$K_a^{e.d1}$	$K_a^{e.d2}$	$K_a^{e.do}$	$K_a^{e...}$

Notes: K = persons (transitions); e = existence; s = survival; o = outside world;
a = age cohort (ATP2)

B. Transition accounts: example for South East, 1966-71

Initial states		Final states: Survival at end in SE	RB	RW	Death in 1966-71 in SE	RB	RW	Totals
Exist 1966	South East	985.5	92.7	101.8	3.7	0.2	0.2	1184.0
	Rest Britain	93.3	2170.4	112.8	0.2	8.4	0.2	2385.3
	Rest World	88.9	77.5	0	0.2	0.1	0	166.8
Totals		1167.8	2340.7	214.6	4.0	8.7	0.4	3736.1

Notes: Age group transition: 20-24 to 25-29. All figures in 1000s.
Source: Rees 1979.

Table 10.4: (continued)

C. Movement accounts: algebraic variables

State before move		State after move: Destinations R1	R2	RW	Death in interval	Totals
Origins	Region 1	R_a^{1}	M_a^{12}	M_a^{10}	D_a^{1}	$P_a^{1.}$
	Region 2	M_a^{21}	R_a^{2}	M_a^{20}	D_a^{2}	$P_a^{2.}$
	Rest World	M_a^{01}	M_a^{02}	0^a	0^a	$M_a^{o.}$
Totals		$P_a^{.1}$	$P_a^{.2}$	$M_a^{.0}$	$D_a^{.}$	$R_a^{..}$

Notes: P = population; M = moves; D = deaths; R = residuals; a = age cohort (ATP2); o = outside world.

D. Movement accounts: example for Greater London, 1976-81

State before move	State after move: Destinations 1976-81 GL	RUK	RW	Death 1976-81	Totals
Greater London	254.1	188.9	52.6	1.7	497.3
Rest of UK	165.1	3061.0	136.4	11.3	3373.9
Rest of World	65.9	101.9	0	0	167.8
Totals	485.1	3351.9	189.0	13.0	4039.0

Notes: Age group transition: 20-24 to 25-29. All figures in 1000s for persons.
Source: Rees 1984b.

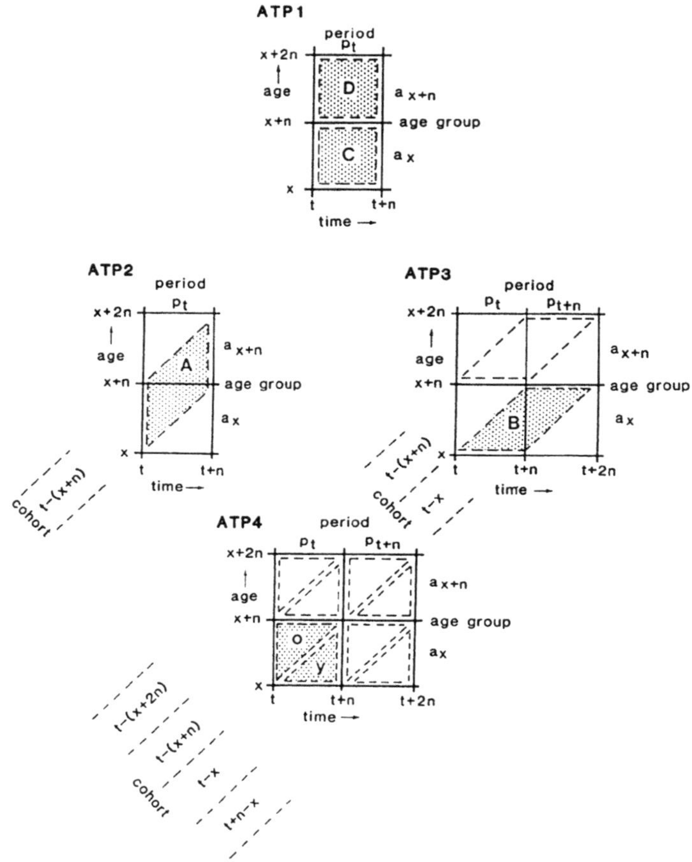

Key
x exact age n age interval t,t+n time points a_x age group starting at age x
p_t period starting at point t ATP age time plan

Fig. 10.3 Age-time observation plans.

gram triangles (for example, o for the older cohort, y for the younger) which can be assembled into either the ATP2 parallelograms (e.g., A) or the ATP3 parallelograms (e.g., B).

In other countries (including the UK) events are classified only by age at time of the event using ATP1 (Figure 10.3). The event data must then be converted to ATP2 or ATP3 by assumption (for example, that events in C are equivalent to half those in D plus half those in C), or by deconsolidation first to ATP4 followed by appropriate re-aggregation (see Rees and Woods, 1985 for examples). The estimations involved in this latter process (ATP1 to ATP4, and hence to ATP2 or ATP3) are often improved by applying rates derived from larger spatial units that contain the study area at the ATP2 or ATP4 level of deconsolidation and constraining the estimates to the observed ATP1 counts (see Rees, 1984b, section 8.2 for an example).

Lessons

In constructing a projective model of a population, the researcher needs to keep in mind the concepts that have been described in this section of the chapter. A clear view must be taken of the system being studied, its boundaries and of the processes by which population enters into and exits from the system. This will be useful even if the available data are far from ideal. Most systems can be described in terms of the 16 life history types or a reduced set clearly related to them, even if population accounts covering the whole system are not constructed.

The researcher should assess which of the possible classifications of a population suit his or her projection problem. Usually age and sex will be used as many other social processes vary in fundamental ways with age and across the sexes, although in many cases age acts only as an intermediate variable for particular life cycle events or its effects are attentuated by contextual variables such as housing tenure, as Courgeau (1984) has elegantly demonstrated with respect to migration. The classifications chosen and their detail will have a profound effect on model design.

Knowledge of the type of transfer data available is essential in model design. Direct measurement of the transition probabilities for discrete time intervals used in many population models is possible only with transition data, although movement data for short time intervals for most processes will provide a good approximation. Conversely, only movement data provides direct measures of the numbers of transfers occurring - transition data will tend to underestimate these. Last migration data, a product of a great many surveys and censuses, falls between the transition and movement stools, and is best converted into the former prior to use in population models.

Finally, it is essential that, before any data are input to population models, the researcher thoroughly understands the age-time plan involved in its collection, or, if survey or life history data for individuals are used, the appropriate age-time plan is employed in producing tabulations for modelling work.

FUNDAMENTAL MODELS

Most population geographers will be familiar with the two most frequently used projective models: the cohort-survival model and the life table or stationary population model. Here attention will be directed at linking these models with the concepts outlined in one or two new ways.

Population Projections and Life Tables

Projective models involve the prediction of how cohorts of the population will evolve in size and characteristics as they age and move forward in time, and how new cohorts will be added to the existing population. It does not matter whether the model is formulated using the population counted in age groups at a point in time in ATP2 or at exact age points over an interval of time in ATP3. Traditionally, the former approach has been used because of the availability of population stocks from censuses taken at one point in time. From many viewpoints knowledge of the second kind of population, in effect, an average population for a period, might be just as useful, but the population stocks would have to be measured from population register records.

A second use of projective models has been to measure duration of life by making some assumptions about the time spent in various states by persons making transitions between them. Traditionally, these life table models have been formulated using ATP3 because actuaries have always made their survival calculations based on a person's exact age. There is no conceptual reason why life tables cannot be developed for age groups using ATP2 or why exact age life tables cannot be developed from data gathered using ATP2 (although Ledent and Rees, 1980, suggest methods for doing this they miss the full implications of their transition models). Age group life tables are less useful than exact age ones because they average over survival chances which vary somewhat within an age group. Survival chances measured in the conventional life table vary with time, but this variation is less marked.

A General Matrix Framework

Leaving choice of age-time plan to one side, dependent as it is on modelling purpose, we propose here a general formulation for multistate projection models that incorporates the points about system structure and type of transfer data made above, benefiting from earlier work by Stone (1975, Chapter VII) and more recent work by Willekens and Drewe (1984) and Keilman (1984) but differing from all of these in several aspects. A system framework intermediate between that of Stone (1975, Table 7.1, p. 42 - in effect, Figure 10.1) and that of Rees and Wilson (1973 - in effect, Table 10.1) is adopted (cf. Table 10.2) to represent the transition (Table 10.5) and movement (Table 10.6) flows that connect opening and closing population stocks. These tables represent the processes of transfer among the states within the system, of transfer out of the system states through death and emigration, and of transfer into the system as a result of immigration and in the case of the lower half of each table (Tables 10.5B, 10.6B) of transfer into the system through birth. A matrix exposition is used because that is the commonest method of presentation used in the literature, although occasional excursions are made into algebraic notation where appropriate.

The symbols in the two tables have the following meanings. The definitions of corresponding elements are listed in parallel order for easy comparison. All terms in the first part of the tables (Tables 10.5A, 10.6A) refer to stocks and flows in an age cohort (ATP2); all terms in the second part of the tables (Tables 10.5B, 10.6B) refer to the first age cohort, labelled o, involving the transition between birth and the first age group.

Table 10.5 Terms

a matrix of transitions between existence and survival states. Persons are classified by state at the start and state at the end of the time interval.

a matrix of transitions between existence and death states. Persons are classified by state at the start of the period and state at death during the interval.

a column vector of transitions from existence states to the outside world, that is, emigrants whether they survive to the end of the period or not.

a column vector of initial populations classified by existence state at the start of the time interval.

a row vector of transitions from the outside world to survival states, or immigrants who survive the time interval in the system of interest.

a row vector of transitions between the outside world and death states in the system or immigrants who die within the system in the time interval.

a scalar, the sum of all immigrants to the system, survivors and non-survivors.

a row vector of final populations classified by survival states at the end of the time interval.

a row vector of deaths, classified by state at death.

a scalar, the sum of all emigrants from the system.

a scalar, the total of all transitions (persons) involved in the system of interest.

Table 10.5: A Demographic Matrix Connecting Opening and Closing Population Stocks in a Time Interval Using Transition Data

A. For existing cohorts (ATP2)

Initial states		Final states: Inside system: Survival at end (s)	Final states: Inside system: Death in interval (d)	Final states: Outside system (emigration) (o)	Totals (opening stocks) (.)
Inside system	Existence at start (e)	$\mathbf{K}_a^{es}$	$\mathbf{K}_a^{ed}$	$\mathbf{k}_a^{eo}$	$\mathbf{k}_a^{e\cdot}$
Outside system	(immigration) (o)	$\mathbf{k}_a^{os}$	$\mathbf{k}_a^{od}$	0	$\mathbf{k}_a^{o\cdot}$
Totals	(closing stocks) (.)	$\mathbf{k}_a^{\cdot s}$	$\mathbf{k}_a^{\cdot d}$	$\mathbf{k}_a^{\cdot o}$	$\mathbf{k}_a^{\cdot\cdot}$

Note: a is the age cohort label (ATP2); a > 1

B. <u>For the Infant Cohorts (ATP2)</u>

Initial states		Final states: Inside system: Survival at end (s)	Final states: Inside system: Death in interval (d)	Final states: Outside system (emigration) (o)	Totals (opening stocks) (.)
Inside system	Birth in interval (o)	$\mathbf{K}_0^{bs}$	$\mathbf{K}_0^{bd}$	$\mathbf{k}_0^{bo}$	$\mathbf{k}_0^{b.}$
Outside system	(immigration) (o)	$\mathbf{k}_0^{os}$	$\mathbf{k}_0^{od}$	0	$k_0^{o.}$
Totals (closing stocks) (.)		$\mathbf{k}_0^{.s}$	$\mathbf{k}_0^{.d}$	$k_0^{.o}$	$k_0^{.o}$

Variables - $\mathbf{K}$: a matrix variable; $\mathbf{k}$: a vector variable; k: a scalar variable.
Subscripts - a: refers to transitions taking place between an age group a_x at time t and the next, a_{x+n}, at time t+n, where n is both the age and time interval (see Figure 10.3, ATP2, area A).
0: refers to transitions taking place between birth, or age 0, during a time interval and the first age group, a_0, at time t+n.
Superscripts identify submatrices by lifestate combination.

Table 10.6: A Demographic Matrix Connecting Opening and Closing Population Stocks in a Time Period Using Movement Data

A. For existing cohorts (ATP2)

Origin states	Destination states: Inside system: Destinations	Destination states: Inside system: Death	Outside system (emigrations)	Totals (opening stocks)
Inside system Origins	$\mathbf{M}_a$	$\mathbf{d}_a$	$\mathbf{e}_a$	$\mathbf{p}_a^I$
Outside system (immigrations)	$\mathbf{i}_a$	0	0	i_a
Totals (closing stocks)	$\mathbf{p}_a^F$	d_a	e_a	t_a

Note: a is the age group label (ATP2); a > 1.

B. For infant cohorts (ATP2)

Origin states	Destination states: Inside system: Destinations	Destination states: Inside system: Death	Outside system (emigrations)	Totals (opening stocks)
Inside system Origins	$\mathbf{M}_o$	$\mathbf{d}_0$	$\mathbf{e}_0$	$\mathbf{b}$
Outside system (immigrations)	$\mathbf{i}_0$	0	0	i_0
Totals (closing stocks)	$\mathbf{p}_0^F$	d_0	e_0	t_0

Variables - **M**: matrix variable; **d**: vector variable; i: a scalar variable.
Subscripts - a refers to movements taking place at age between an age group a_x at time t and the next, a_{x+n}, at time t+n, where n is both the age and time interval (see Figure 10.3, ATP2, area A).
0 refers to movements taking place between birth, or age 0, during a time interval and the first age group, a_0, at time t+n.

Table 10.6 terms

M_a a matrix of moves between origin and destination states. Moves are classified by state before the move and state after the move.

d_a a column vector of deaths classified by origin state, the state which a person leaves on dying.

e_a a column vector of emigrations classified by origin state, the state a person leaves on emigrating.

p_a^I a column vector of initial populations classified by origin state at the start of the time interval.

i_a a row vector of immigrations classified by destination state, the state a person enters on immigration.

i_a a scalar, the sum of all immigrations into the system.

p_a^F a row vector of final populations, classified by destination state at the end of the time interval.

e_a a scalar, the sum of all emigrations from the system.

t_a a scalar, the total of all moves involving the system of interest.

The terms for the infant cohort are all equivalent, except that birth states are substituted for existence states and opening stocks:

k_0^b a column vector of total births classified by state at birth, summed over mothers' ages at maternity (Table 10.5B term).

b a column vector of births classified by origin state at birth, summed over mothers' ages at maternity.

Note that, apart from the opening and closing population stocks, also directly equivalent are the two deaths vectors, $k_a.^d$ and d_a, although they appear in different locations in the transition and movement accounts tables.

A Projection Model

Given these definitions, we can now develop population projection models incorporating outflow coefficients that "survive" initial population stocks for both transition and movement cases. Survival equations are needed for both existing and infant cohorts, and a fertility model is needed to generate the opening stocks (births) of the infant cohort from the populations at risk in the existing cohort. The models are specified with exogeneous inputs (immigrants or immigrations).

The closing population stocks are given by, in the transition case,

$$\underset{\sim}{k}_a{}^{.s} = \underset{\sim}{1}\ \underset{\sim}{K}_a{}^{es} + \underset{\sim}{k}_a{}^{os} \tag{10.1}$$

$$\underset{\sim}{k}_o{}^{.s} = \underset{\sim}{1}\ \underset{\sim}{K}_o{}^{bs} + \underset{\sim}{k}_o{}^{bs} \tag{10.2}$$

and in the movement case by

$$\underset{\sim}{p}_a(t+1) = \underset{\sim}{1}\ \underset{\sim}{M}_a + \underset{\sim}{i}_a \tag{10.3}$$

$$\underset{\sim}{p}_o(t+1) = \underset{\sim}{1}\ \underset{\sim}{M}_o + \underset{\sim}{i}_o \tag{10.4}$$

where 1 is a row vector of ones.

To link closing population stocks to initial we need to define outflow coefficients. In the transition case each element in the $K_a{}^{es}$ matrix is divided by its corresponding row total, initial populations for existing cohorts

$$\underset{\sim}{C}_a{}^{es} = (\hat{\underset{\sim}{k}}_a{}^{e.})^{-1}\ \underset{\sim}{K}_a{}^{es} \tag{10.5}$$

and for the infant cohorts, births

$$\underset{\sim}{C}_a{}^{bs} = (\hat{\underset{\sim}{k}}_o{}^{b.})^{-1}\ \underset{\sim}{K}_o{}^{bs} \tag{10.6}$$

where the circumflex indicates that the column vectors have been spread out to form a diagonal matrix so that, for example,

$$(\hat{\underset{\sim}{k}}_a{}^{e.})^{-1} = \begin{vmatrix} \frac{1}{K_a{}^{e1..}} & 0 & \cdots & 0 \\ 0 & \frac{1}{K_a{}^{e2..}} & \cdots & 0 \\ \vdots & \vdots & \ddots & \vdots \\ 0 & 0 & \cdots & \frac{1}{K_a{}^{en..}} \end{vmatrix} \tag{10.7}$$

The outflow coefficients from the Table 10.4B example would be, for instance, 98.5/1184.0 or .832 for the flow of surviving stayers within the South East region and 92.7/1184.0 or .078 for migrants (surviving) from the South East to the Rest of Britain. These coefficients are transition probabilities.

If the $K_a{}^{es}$ and $K_a{}^{bs}$ matrices in equations (10.1) and (10.2) are substituted for by expressions derived from equations (10.5) and (10.6) respectively, we obtain projection equations for the closing population stocks in existing cohorts

$$\begin{aligned} \underset{\sim}{k}_a{}^{.s} &= \underset{\sim}{1}\ \underset{\sim}{k}_a{}^{e.}\ \underset{\sim}{C}_a{}^{es} + \underset{\sim}{k}_a{}^{os} \\ &= \underset{\sim}{k}_a{}^{e.\prime}\ \underset{\sim}{C}_a{}^{es} + \underset{\sim}{k}_a{}^{os} \end{aligned} \tag{10.8}$$

and for the infant cohort

$$\underset{\sim}{k}_0{}^{.s} = \underset{\sim}{1}\, \underset{\sim}{k}_0{}^{b.}\, \underset{\sim}{C}_0{}^{bs} + \underset{\sim}{k}_0{}^{os}$$

$$= \underset{\sim}{k}_0{}^{b.\prime}\, \underset{\sim}{C}_0{}^{bs} + \underset{\sim}{k}_0{}^{os} \qquad (10.9)$$

where the prime indicates transposition making row vectors of initial populations or births totals.

Outflow coefficients computed by dividing the flow term by the initial population are appropriate for the transition case because only the initial population contributes to the flow. In the movement case, however, the contributing populations are more varied: a migration in the 1976-81 period from Greater London to the Rest of the UK could be made by a person in the Greater London population in 1976, or by a person who was in another region initially and who migrated to Greater London. The proper population at risk is the average population over the period. Normally, an arithmetic average population is adopted with the assumption that the population takes a linear path between start and end of the time interval.

The outflow coefficients in the movement case are, using algebraic notation of Table 10.4C,

$$m_a{}^{ij} = M_a{}^{ij}/.5\ (P_a{}^{i.} + P_a{}^{.i})$$

$$m_a{}^{io} = M_a{}^{io}/.5\ (P_a{}^{i.} + P_a{}^{.i})$$

$$d_a{}^{i} = D_a{}^{i}\ /.5\ (P_a{}^{i.} + P_a{}^{.i}) \qquad (10.10)$$

where lower case letters are used for these occurrence-exposure rates of inter-state movement within the system, of emigration to the outside world, and of death. These rates cannot be used directly in the projection equations because the rate denominators are not the initial populations.

Drawing on derivations by Rogers and Ledent (1976), Willekens and Rogers (1978), Rees (1983b), Willekens and Drewe (1984) and Keilman (1984), this problem can be solved in the following way. First, we use algebraic notation and then generalise this result to a matrix equation. The closing population stocks in an age group (ATP2) are given by

$$P_a{}^{.i} = P_a{}^{i.} - (\sum_{j\neq i} M_a{}^{ij} + D_a{}^{i} + M_a{}^{io})$$

$$+ \sum_{j\neq i} M_a{}^{ji} + M_a{}^{oi} \qquad (10.11)$$

Substituting for the internal migration, deaths and emigration terms from expressions derived from equation set (10.10), we obtain

$$P_a{\cdot}^i = P_a{}^{i\cdot} - \left(\sum_{j\neq i} m_a{}^{ij} + d_a{}^i + m_a{}^{io}\right)(.5(P_a{}^{i\cdot} + P_a{\cdot}^i)$$

$$+ \sum_{j\neq i} m_a{}^{ji}\ (.5(P_a{}^{j\cdot} + P_a{\cdot}^j) + M_a{}^{oi} \qquad (10.12)$$

which is repeated for all states i = 1 to N.

These N equations can be rewritten in matrix form, using the notation of Table 10.6

$$\underset{\sim}{p}_a{}^{F'} = [\underset{\sim}{I} + .5\ \underset{\sim}{M}_a]^{-1}\ [\underset{\sim}{I} - .5\ \underset{\sim}{M}_a]\ \underset{\sim}{p}_a{}^I$$

$$+ [\underset{\sim}{I} + .5\ \underset{\sim}{M}_a]^{-1}\ \underset{\sim}{i}_a \qquad (10.13)$$

where M_a is defined as

$$\underset{\sim}{M}_a = \begin{vmatrix} \left(\sum_{j\neq 1} m_a{}^{1j} + d_a{}^1 + m_a{}^{1o}\right) & -m_a{}^{21} & \cdots & -m_a{}^{n1} \\ -m_a{}^{12} & \left(\sum_{j\neq 2} m_a{}^{2j} + d_a{}^2 + m_a{}^{2o}\right) & \cdots & -m_a{}^{n2} \\ \vdots & \vdots & \ddots & \vdots \\ -m_a{}^{1n} & -m_a{}^{2n} & \cdots & \left(\sum_{j\neq n} m_a{}^{nj} + d_a{}^n + m_a{}^{no}\right) \end{vmatrix}$$

Then, if we define the matrix S_a thus

$$\underset{\sim}{S}_a = [\underset{\sim}{I} + .5\ \underset{\sim}{M}_a]^{-1}\ [\underset{\sim}{I} - .5\ \underset{\sim}{M}_a] \qquad (10.15)$$

then the population projection model of equation (10.13) may be written

$$\underset{\sim}{p}_a{}^{F'} = \underset{\sim}{S}_a\ \underset{\sim}{p}_a{}^I + .5\ [\underset{\sim}{I} + \underset{\sim}{S}_a]\ \underset{\sim}{i}_a \qquad (10.16)$$

The S matrix is the matrix of outflow coefficients that transform the initial or opening stocks into the final. Both Willekens and Drewe (1984) and Keilman (1984) interpret these coefficients as

> "period-cohort transition probabilities... that a person... classified in status i at time t will 'survive' until t+1... and at that time will occupy status j." [Keilman, 1984, p.11].

There are several arguments against this view (see Rees, 1985, for a detailed case) but the most fundamental is that no amount of mathematical manipulation will convert movement rates into transition probabilities unless information on the relationship of moves to transitions is introduced.

For the infant cohort (those born in the time interval) the movement rates are defined as

$$m_0^{ij} = M_0^{ij}/.5\ (B_{.}^{\ i} + P_0^{\ .i})$$

$$m_0^{io} = M_0^{io}/.5\ (B_{.}^{\ i} + P_0^{\ .i})$$

$$d_0^{\ i} = D_0^{\ i}/.5\ \ (B_{.}^{\ i} + P_0^{\ .i}) \qquad (10.16)$$

where $B_{.}^{\ i}$ refers to births in region i (summed over all ages of mother). Defining M_0 in the same way as M_a, we obtain the system of projection equations for the infant cohort

$$\underset{\sim}{p}_0'(t+1) = [\underset{\sim}{I} + .5\ \underset{\sim}{M}_0]^{-1}\ [\underset{\sim}{I} - .5\ \underset{\sim}{M}_a]\ b_{.}$$

$$+ [\underset{\sim}{I} + .5\ \underset{\sim}{M}_a]^{-1}\ \underset{\sim}{i}_o$$

$$= \underset{\sim}{S}_0\ \underset{\sim}{b}_{.} + .5\ [\underset{\sim}{I} + \underset{\sim}{S}_0]\ \underset{\sim}{i}_0 \qquad (10.17)$$

Willekens and Drewe (1984) and Keilman (1984) have different equations for the infant cohort in which the population at risk for occurrence-exposure rates are inconsistently defined.

Modifications of the equation sets for existing cohorts a = 1 to A are suggested for the last, the Ath, by Willekens and Drewe (1984) and Keilman (1984). In fact, such modifications are unnecessary if the adjustment (adding together the final populations of the A-1th and Ath cohort) is postponed until the conversion of final populations for one period into the initial populations of the next. In the transition case this is accomplished thus, attaching time interval labels to the initial and final populations (q is the current period, q-1 the previous period)

$$\underset{\sim}{k}_1^{\ e.}(q) = \underset{\sim}{k}_0^{\ .s}(q-1)' \qquad \text{for age group 1}$$

$$\underset{\sim}{k}_a^{\ e.}(q) = \underset{\sim}{k}_{a-1}^{\ .s}(q-1)' \qquad \text{for age groups 2 to A-1}$$

and for the last age group A

$$\underset{\sim}{k}_A^{\ e.}(q) = \underset{\sim}{k}_{A-1}^{\ .s}(q-1)' + \underset{\sim}{k}_A^{\ .s}(q-1)' \qquad (10.18)$$

and in the movement case

$$\underset{\sim}{p}_1^{\ I}(q) = \underset{\sim}{p}_0^{\ F}(q-1)' \qquad \text{for age group 1}$$

$$\underset{\sim}{p}_a{}^I(q) = \underset{\sim}{p}_{a-1}{}^F(q-1)' \quad \text{for age groups 2 to A-1}$$

and for the last age group A

$$\underset{\sim}{p}_A{}^I(q) = \underset{\sim}{p}_{A-1}{}^F(q-1)' + \underset{\sim}{p}_A{}^F(q-1) \tag{10.19}$$

Initial populations for the new period in the last age cohort A are given by adding together the projections of the last cohort and the last but one: for example, the population that survives from 85+ to 90+ together with that which survives from 80-84 to 85-89 make up the initial population for the next period of the 85 and over age group.

The final step in the projection model is to insert between the existing cohort and infant cohort projection equations a set of equations for generating the births totals that enter the infant cohort projection. The initial stocks for the infant cohort are normally the sum of births to mothers of different ages over the fertile age range, that is,

$$\underset{\sim}{k}_0{}^{b\cdot} = \sum_{a=a_1}^{a_2} \underset{\sim}{k}_0{}^{b\cdot}(a) \tag{10.20}$$

or

$$\underset{\sim}{b} = \sum_{a=a_1}^{a_2} \underset{\sim}{b}(a) \tag{10.21}$$

where (a) refers to the age of mother (ATP2).

In the projection model based on transition data, the next step is to make the births a function of the initial populations, just as the exist-survive transitions were so made. This assumes that the births are classified by initial location of parent or mother. This assumption can be dropped later. Fertility rates can be defined as

$$\underset{\sim}{F}_a = \hat{\underset{\sim}{k}}_a{}^{e.-1} \, \underset{\sim}{k}_0{}^{b\cdot}(a) \tag{10.22}$$

that is, births divided by the initial populations, where F_a is a diagonal matrix with fertility rates on the principal diagonal. Unfortunately, very few systems of recording vital events classify births by mother's previous life history and it would be difficult to do so. Fertility rates must be defined by dividing by the average population in the same fashion as the movement projection model. Individual fertility rates are defined in the transition case as

$$f_a^{\,i} = K_a^{\,bi..}/.5\,(K_a^{\,ei..} + K_a^{\,..si}) \tag{10.23}$$

and in the movement case as

$$f_a^{\,i} = B_a^{\,i}/.5\,(P_a^{\,i.} + P_a^{\,.i}) \tag{10.24}$$

The population at risk (PAR), expressed as a column vector, using the linear assumption, are respectively in the transition case

$$\begin{aligned} \underset{\sim}{PAR}_a &= .5\,[\underset{\sim}{k}_a^{\,.s'} + \underset{\sim}{k}_a^{\,e.}] \\ &= .5\,[[\underset{\sim}{C}_a^{\,es'}\,\underset{\sim}{k}_a^{\,e.} + \underset{\sim}{k}_a^{\,os}] + \underset{\sim}{k}_a^{\,e.}] \\ &= .5\,[\underset{\sim}{I} + \underset{\sim}{C}_a^{\,es'}]\,\underset{\sim}{k}_a^{\,e.} + .5\,\underset{\sim}{k}_a^{\,os} \end{aligned} \tag{10.25}$$

and in the movement case

$$\begin{aligned} \underset{\sim}{PAR}_a &= .5\,[\underset{\sim}{p}_a^{\,F'} + \underset{\sim}{p}_a^{\,I}] \\ &= .5\,[[\underset{\sim}{S}_a\,\underset{\sim}{p}_a^{\,I} + .5\,[\underset{\sim}{I} + \underset{\sim}{S}_a]\,\underset{\sim}{i}_a] + \underset{\sim}{p}_a^{\,I}] \\ &= .5\,[\underset{\sim}{I} + \underset{\sim}{S}_a\,\underset{\sim}{p}_a^{\,I}] + .25\,[\underset{\sim}{I} + \underset{\sim}{S}_a]\,\underset{\sim}{i}_a \end{aligned} \tag{10.26}$$

Then, if a female dominant fertility model is adopted, births are generated, summed and sexed thus

$$\underset{\sim}{k}_0^{\,b.x} = z^x \sum_{a=a_1}^{a_2} \underset{\sim}{F}_a\,\underset{\sim}{PAR}_a^{\,f} \tag{10.27}$$

$$\underset{\sim}{b}^x = z^x \sum_{a=a_1}^{a_2} \underset{\sim}{F}_a\,\underset{\sim}{PAR}_a^{\,f} \tag{10.28}$$

where a superscript f refers to the female population at risk, where F_a is a diagonal matrix of fertility rates defined as in equations (10.23) or (10.24), z^x is the proportion of all births of sex x, and a_1 and a_2 are the youngest and oldest fertile cohorts respectively. The birth totals are entered in equations (10.9) and (10.17) respectively to obtain the survived population in the first age group by system states at the end of the time interval.

The projection models are implemented by recursion over successive time periods using either fixed coefficients and immigration vectors (stability analysis - as explored by Alexander, 1983) or variable coefficients and immigration vectors.

SIMPLIFICATIONS AND COMPLICATIONS

Some brief remarks are made in this concluding section on some of the problems faced in implementing the projection models and some of the solutions suggested.

Flexible Models

In the previous section a matrix representation of population projection models was used as this is the commonest method of exposition used. However, the matrix model has a rather rigid structure for many purposes, and an alternative is to develop a more flexible modelling system tied to suitable software. Full details of the modelling system are given in Rees (1981) for transition models and Rees (1984b) for movement models. A brief sketch of the principles underlying these modelling systems is given here.

The basic idea is to estimate each of the components of the accounts matrices (the individual cells in Tables 10.5 and 10.6) separately, and to provide a set of four choices for each component. The first choice is to enter flow or stock data for the component. The second choice is to use the flow or stock data of the previous modelling period. The third choice is to enter rate data for the component along with a choice of appropriate population at risk (initial or final populations, average populations or multistate populations at risk). The fourth choice is to employ the rate data for the previous period. There are also some other choices which enable the researcher to choose between transmission models (all those in the previous section are transmission models) and admission models, and between various methods of adjusting the accounts matrices to ensure consistency.

With such modelling systems it is possible to explore a variety of possible ways of closing the system and of estimating the base period accounts matrices (Rees, 1983b). Many different kinds of data can be used.

Dealing with a Large State-Space

The presentation above deliberately did not specify what states were used to classify the population within the system in order to retain general relevance. However, as was pointed out earlier, adoption of more than one or two dimensions in the classification leads to difficulties of unreliable, sparse matrices.

Three approaches to this problem can be identified: the first is to decompose or aggregate the system spatially, the second is to substitute rates generated from parameterized models for as many of the model components as possible, and the third is to restructure the models as microsimulation models at the individual scale using linked chains of probability equations relating selected sets of characteristics for an application.

For example, Rogers (1976) has shown that a set of models involving a region and the rest of the country provide a good approximation of the results of a multiregional cohort survival model. Masser (1976) has shown that hierarchical models of migration capture most detail along with the grosser interregional flows.

Rogers has recently summarised his extensive work on modelling schedules of age classified migration, fertility, first marriage and divorce in an illustrative projection for a two region-four marital statuses system for Sweden (Rogers and Planck, 1984). Work on the application of spatial interaction models in which flows are represented by a simpler set of origin and destination terms and cost functions continues apace (Ledent, 1981, 1985; Stillwell, 1983) together with work on related estimation, statistical and log-linear models (Willekens, 1982, 1983a, 1983b). The integration of such models in a multistate projection framework has still to be accomplished.

Microsimulation methods (Clarke, Keys and Williams, 1981) have great potential in enabling researchers to deal with household formation, transition and dissolution more effectively (Clarke, 1985) as the complicated event sequences and interactions are not conveniently represented in aggregate models of the type outlined. Microsimulation methods are also likely to be useful in solving a problem of classification peculiar to systems in which an individual's characteristics are inherited. An example would be the ethnic intermixture that results from cross-ethnic marriage and fertility: in theory, the possible combinations square with each generation, and to date the problem has been avoided in ethnic population models (see Immigrant Statistics Unit, 1979; and Ram and Ree, 1985 for a discussion).

Demography is not enough

As Stone (1975) makes very clear in his review, demographic variables alone are insufficient for a proper understanding of planned systems. Population models merely supply possible participants in systems; the decision makers control the activities and a new class of models needs to be used that look at the costs of activities and supply-demand relationships.

Concluding Remarks

Although this chapter takes a partial view of population models, the concepts discussed and the models explained do have considerable value. It was shown that it would be perilous to embark on any exercise in population modelling without a clear idea of the nature of the system being modelled and the kind of population states or classifications necessary for the problem in hand. It would also be dangerous not to give careful consideration to the type of transfer data that could be collected for the system under study. It was demonstrated that population models have to be designed with the characteristics of the transfer data being used firmly in mind. If these lessons are applied in the reader's future work, then the mistakes made by practitioners in the course of arriving at the understandings embodied in this chapter will not have been made in vain.

REFERENCES

Alexander, S. (1983) 'A Model of Population Change with New and Return Migration', Environment and Planning A, 15, 1231-1257

Brown, A. and Fox, J. (1984) 'OPCS Longitudinal Study: Ten Years On', Population Trends, 37, 20-22

Clarke, M. (1985) 'Demographic Processes and Household Dynamics: A Micro-Simulation Approach', in R.I. Wood and P.H. Rees, (eds.), Population Structure and Models, George Allen and Unwin, London

Clarke, M., Keys, P. and Williams, H.C.W.L. (1981) 'Micro-Simulation' in N. Wrigley and R.J. Bennett (eds.), Quantitative Geography, Routledge and Kegan Paul, London

Courgeau, D. (1980) L'Analyse Quantitative des Migrations Humaines, Masson, Paris

Courgeau, D. (1984) 'Relations entre Cycle de Vie et Migrations'. (Relationships between the Life Cycle and Migration), Population, 39, 483-514

Keilman, N. (1984) 'Internal and External Consistency in Multi-Dimensional Population Projection Models'. Working Paper No.46, Netherlands Interuniversity Demographic Institute, Voorburg, The Netherlands

Ledent, J. (1980) 'Multistate Life Tables: Movement Versus Transition Perspectives', Environment and Planning A, 12, 533-562

Ledent, J. (1981) 'On the Relationship between Alonso's Theory of Movement and Wilson's Family of Spatial Intersection Models', Environment and Planning A, 13, 217-224

Ledent, J. (1985) 'The Doubly Constrained Model of Spatial Interaction: A More General Formulation', Environment and Planning A, 17, 253-262

Ledent, J. and Rees, P.H. (1980) 'Choices in the Construction of Multiregional Life Tables'. Working Paper WP-80-173, International Institute of Applied Systems Analysis, Laxenburg, Austria

Masser, I. (1976) 'The Design of Spatial Systems for Internal Migration Analysis', Regional Studies, 10, 39-52

Ram, S. and Rees, P.H. (1985) A Spatial Demographic Analysis of Indians in Bradford. Working Paper 434, School of Geography, University of Leeds

Rees, P.H. (1979) 'Multiregional Population Analysis: The Accounting Approach', Sistemi Urbani, 1, 3-32

Rees, P.H. (1981) Accounts Based Models for Multiregional Population Analysis: Methods, Program and User's Manual. Working Paper 295/Computer Manual 13, School of Geography, University of Leeds

Rees, P.H. (1983a) 'Multiregional Mathematical Demography: Themes and Issues', Environment and Planning A, 15, 1571-1583

Rees, P.H. (1983b) 'Choices in the Construction of Regional Population Projections'. Working Paper 378, School of Geography, University of Leeds. Reproduced in R.I. Woods and P.H. Rees (eds.), Population Structure and Models, George Allen and Unwin, London

Rees, P.H. (1984a) 'Does it Really Matter which Migration Data you Use in a Population Model?' Working Paper 383, School of Geography, University of Leeds. Forthcoming in P.E. White and G.A. van der Knaap (eds.), Contemporary Issues in Migration, Geobooks, Norwich

Rees, P.H. (1984b) Spatial Population Analysis using Movement Data and Accounting Methods: Theory, Models, the 'MOVE' Program and Examples. Working Paper 404/Computer Manual 23, School of Geography, University of Leeds

Rees, P.H. (1985) 'Developments in the Modelling of Spatial Populations', in R.I. Woods and P.H. Rees (eds.), Population Structure and Models, George Allen and Unwin, London

Rees, P.H. and Wilson, A.G. (1973) 'Accounts and Models for Spatial Population Analysis. 1: Aggregate Population', Environment and Planning, 5, 61-90

Rogers, A. (1976) 'Shrinking Large-Scale Population Projection Models by Aggregation and Decomposition', Environment and Planning A, 8, 515-541

Rogers, A. and Ledent, J. (1976) 'Increment-Decrement Life Tables: A Comment', Demography, 13, 287-290

Rogers, A. and Planck, F. (1984) Parameterized Multistate Population Projections. Working Paper 84-1, Population Program, Institute of Behavioral Science, University of Colorado, Boulder

Stillwell, J.C.H. (1983) 'Migration between Metropolitan and Non-Metropolitan Regions in the UK'. Working Paper 367, School of Geography, University of Leeds, Leeds. Reproduced in R.I. Woods and P.H. Rees (eds.), Population Struction and Models, George Allen and Unwin, London

Stone, R. (1971) Demographic Accounting and Model Building, OECD, Paris

Stone, R. (1975) Towards a System of Social and Demographic Statistics, Department of Economic and Social Affairs, Studies in Methods, Series F. no.18, United Nations, New York

Willekens, F. (1982) 'Multidimensional Population Analysis with Incomplete Data' in K. Land and A. Rogers (eds.), Multidimensional Mathematical Demography, Academic Press, New York, pp. 43-111

Willekens, F. (1983a) 'Specification and Calibration of Spatial Interaction Models', Tijdschrift voor Economische en Sociale Geografie, 74, 239-252

Willekens, F. (1983b) 'Log-Linear Modelling of Spatial Interaction', Papers of the Regional Science Association, 52, 187-205

Willekens, F. and Baydar, N. (1983) 'Forecasting Place-to-Place Migration with Generalized Linear Models (GLM) - An Application to Urbanization in the Netherlands'. Working Paper no.42, Netherlands Interuniversity Demographic Institute, Voorburg, The Netherlands. Reproduced in R.I. Woods and P.H. Rees (eds.), Population Structure and Models, George Allen and Unwin, London

Willekens, F. and Drewe, P. (1984) 'A Multiregional Model for Regional Demographic Projection' in H. Ter Heide and F. Willekens (eds.), Demographic Research and Spatial Policy, Academic Press, London, pp. 309-334

Willekens, F. and Rogers, A. (1978) Spatial Population Analysis: Methods and Computer Programs, Research Report RR-78-18, International Institute for Applied Systems Analysis, Laxenburg, Austria

NOTES ON CONTRIBUTORS

Dr M. Cadwallader, Department of Geography, University of Wisconsin, Madison, USA

Dr J. Coward, Department of Geography, University of Ulster, Northern Ireland

Dr P.H. Curson, School of Earth Sciences, Macquarie University, Australia

Dr J.C. Dewdney, Department of Geography, University of Durham, England

Dr A.J. Fielding, Department of Geography, University of Sussex, England

Professor J. Gugler, Department of Sociology, University of Connecticut, USA

Dr M. Pacione, Department of Geography, University of Strathclyde, Glasgow, Scotland

Dr P. Rees, School of Geography, University of Leeds, England

Professor D. Rhind, Department of Geography, Birkbeck College, University of London, England

Dr J. Salt, Department of Geography, University College London, England

Mr I. Thomas, School of Development Studies, University of East Anglia, Norwich, England

Dr R.I. Woods, Department of Geography, University of Sheffield, England

INDEX

For Product Safety Concerns and Information please contact our EU representative GPSR@taylorandfrancis.com
Taylor & Francis Verlag GmbH, Kaufingerstraße 24, 80331 München, Germany

www.ingramcontent.com/pod-product-compliance
Ingram Content Group UK Ltd.
Pitfield, Milton Keynes, MK11 3LW, UK
UKHW022111190926
475702UK00013B/1380
* 9 7 8 0 4 1 5 6 1 6 1 2 6 *